Learner-Centered Astronomy Teaching:
STRATEGIES FOR ASTRO 101

Timothy F. Slater
University of Arizona

Jeffrey P. Adams
Montana State University

ei PRENTICE HALL SERIES IN EDUCATIONAL INNOVATION

Prentice
Hall

Pearson Education, Inc.
Upper Saddle River, New Jersey 07458

Acquisitions Editor: *Erik Fahlgren*
Production Editor: *Kim Dellas*
Art Director: *Jayne Conte*
Cover Design: *DeFranco Design*
Manufacturing Manager: *Trudy Pisciotti*
Manufacturing Buyer: *Alan Fischer*
Senior Marketing Manager: *Mark Phaltzgraff*
Assistant Editor: *Christian Botting*
Copy Editor: *Patricia Daly*

©2003 by Pearson Education, Inc.
Pearson Education, Inc.
Upper Saddle River, New Jersey 07458

Printed in the United States of America
10 9 8 7 6 5 4 3

ISBN 0-13-046630-1

Pearson Education LTD., *London*
Pearson Education Australia PTY. Limited, *Sydney*
Pearson Education Singapore, Pte. Ltd.
Pearson Education North Asia Ltd, *Hong Kong*
Pearson Education Canada, Ltd., *Toronto*
Pearson Educacion de Mexico, S.A. de C.V.
Pearson Education—Japan, *Tokyo*
Pearson Education Malaysia, Pte. Ltd.

Contents

Foreword

by
Michael Zeilik

The first day of class, fall semester! You, the instructor, arrive early to observe your students amble into a concrete, tiered lecture hall with seats bolted to the floor. You expect some 300 to arrive, most on your class list, a few not. They are nonscience majors. They have weak analytical skills (graphing for most is akin to going to the dentist). Yet, the electricity of a startup crosses the room and lights up the minds of the student—at least today.

You now face the teacher's eternal dilemma: What do I do in class today? And the next class. And the next. If you have taught as many years as I have (about 30), you know that the electricity will disperse in few weeks in a "standard-model" course. Let me give you an example of "standard-model" classes.

BEFORE
Class #1: A multimedia "tour of the cosmos." Great fun for everyone!

Class#2: Prof. Z opens with the naked-eye sky. He tries to break the passivity by asking if anyone has questions. None are asked. He then queries the students about what they observe in the sky. Silence for a while, but Prof. Z. knows to wait it out. A few shy hands go up; students start with the sun, moon, and stars. Prof. Z. then steers the class into the motions of these objects. At the blackboard, he divides the responses up into motions relative to the horizon and those relative to the stars. A few students state that the moon goes around the zodiac once in a year. Prof. Z. has expected this response and compares the sun's motion to that of the moon. The whole-class discussion becomes lively and then falters with retrograde motion. Prof. Z. anticipates this halt and turns to his portable Mac with

---

[1] Copyright © 2002, Michael Zeilik

Voyager II software. The classroom has a large-screen video projector; all 300 students can see the display. The program is scripted so that it starts with the simplest motions and finishes with the most complex ones. Prof. Z. feels that some students have learned from the visualization. He then summarizes the main motions in a table shown with an overheard projector. The class period is almost over; he asks for questions. A student asks about the first test. He feels that it went pretty well; they have been quiet and attentive. Between the class and the textbook, he believes that most students should get the basics.

Class #3: A short multiple-choice quiz. The results are disappointing; class average is about 60%. Some 45% respond that the moon moves through the zodiac once in a year. Prof. Z. sighs; over all these years, same result, even though he works hard to make these points clear.

Of course, Prof. Z. is me. And I used to do the same thing over and over again, expecting different results but getting more of the same. Then one semester I gave to a "good" class what I thought was a very simple final exam that focused on the big picture of astronomy. The class bombed it; the average was less than 50%. That roused me to action! Desperately seeking solutions, I started reading about undergraduate education and its success. I found a depressing picture. Less than half of full-time students graduate in five years. In science and related disciplines, students who major and those who eventually switch out complain of poor pedagogy in their (usually large) introductory courses. Standard-model courses only reach about 5 to 10% of today's students, no matter how excellent the instructor as a lecturer.

I also found a silver lining. Cognitive sciences have given us new insights on how people learn. Disciplinary-based research in physics and astronomy provide evidence of how to reform our courses. When combined, they grant us the wisdom for reforming our courses. The basic rule: student-centered instruction that actively engages their minds, with lots of peer interaction, plenty of formative assessment integrated with instruction, and a focus on concepts.

Fair enough, but *how*? Tim Slater and Jeff Adams have brought those evidence-based *hows* together in this book, the first compilation of its kind. Here you will find tools—not tricks!—to make your astronomy classes more effective and exciting, to keep the electricity flowing throughout the semester. You will find a diagnostic survey to assess conceptual understanding, and an attitude survey to probe changes in students' attitudes toward astronomy in particular and science in general.

Adams and Slater write from *their* experiences in reforming *their* courses. I did so, too, and here are the results.

AFTER

Class #1: No multimedia tour. Prof. Z. states that this class will be very different from other large classes at the university. He requests that the students form into small groups. They do so, obviously puzzled. He then instructs them to introduce themselves. Silence at first, then the start of a buzz that quickly rises in level. As it tapers, Prof. Z. then asks the groups to come up with a written consensus to the question: "What is astronomy?" They dive in, and he wanders around the class introducing himself to students and asking their names. The buzz reaches a high level of intensity. Prof. Z. walks to the front of the class and initiates an all-class discussion. He asks a spokesperson for some groups to give their responses, and he comments briefly on each. All groups hand in consensus reports, which he will read that evening. He explains the purpose and function of cooperative learning groups in the class, and identifies the social roles involved. He then gives the class a multiple-choice misconceptions assessment, a concept map assessment, and a student background form to be filled in outside of class.

Class #2: Prof. Z. reads the best of the responses from the class #1 activity. He has a "Q&A" as a forum for questions; and few are asked. He then uses *Voyager II* and the large-screen video projector to simulate naked-eye motions in the sky. He asks the class to reform their learning teams and reminds them of their roles as cooperative learners. He hands out the first activity. It relies on graphs of planetary motions in the zodiac. He knows from the misconceptions pretest that many students will find this difficult, so as he tours the class he makes sure that students read the graphs correctly. Many struggle, as if they had never seen a graph before. Prof. Z. tries to visit all groups in about 30 to 40 minutes. He then closes the class with a discussion of the activity and a concept map on heavenly motions. Written consensus reports are handed in.

Class #3: A short multiple-choice quiz. Students work on them individually, then form into their groups for discussion. The classroom jumps to life as the students argue their answers. Class average is about 70%. Some 85% of the students correctly answer that the moon takes a month to complete one circuit of the zodiac. Yes! They are getting it.

That is your challenge, and this book will help you confront that challenge!

Preface

Imagine, if you will, an introductory astronomy classroom where all students pay attention during your lectures, come to class having studied the assigned reading, and have thoughtful and insightful questions ready to pose. Because of your carefully planned sequence of topics, they understand the big ideas in astronomy. Because of your instruction, they can answer challenging questions. Because of your contagious enthusiasm, they adopt a positive view about science even though they are predominantly nonscience majors. Is this pure fantasy? Honestly, it might be—nevertheless, some aspects are certainly achievable. But, just what exactly do busy faculty have to do to make progress in this direction?

Our goal in writing this book is to present a mix of tried-and-true teaching strategies, results from research in teaching and learning, and some of our own "in the trenches" experiences to help faculty interested in engaging in a process of continual improvement designed to enhance student outcomes and teacher satisfaction. Some of the ideas presented here will definitely work in your course while others will need some, possibly significant, adaptation. To be clear, we are not advocating that all of these ideas must be, or even should be, implemented uncritically. However, we firmly believe that reflecting on how your course might look different, and how your students could be different as a result of your course, is a healthy exercise that too few faculty take time to pursue.

Just so you know where we are coming from, we both teach in large-enrollment environments in large universities where our students are nonscience majors, think of themselves as predominantly math and science phobic, and generally enroll for the express purpose of fulfilling a general science requirement. In other words, astronomy seemed like the most painless of the choices available to them. Accordingly, the focus of this book is on this large-enrollment environment; however, it is our experience that most ideas that work at all in very large classes work even better when adapted to smaller-enrollment courses.

We both actively conduct research on the teaching and learning of physics and astronomy and allocate considerable amounts of our time to curriculum development. Our thinking about teaching and learning in astronomy has been profoundly influenced by the results of physics education research, which has repeatedly demonstrated that students are often able to convince their professors that they understand a concept when they actually have only a superficial knowledge of it. Further, we have adopted a view that most of our students are not like us—they do not learn best through lectures, no matter how clearly presented, and the questions that most interest us as scientists are not always the same questions that engage nonscience majors. You are certainly welcome to disagree with these perspectives at a variety of levels—certainly even we do not adhere to them 100% of the time—but we state them so that you can understand the nature of our commentary.

We would like to thank Dr. Dana Lehr and Dr. Christopher Sirola for their careful review of the first draft of this manuscript and Patricia Daly for her careful copyediting of the final version. Any remaining errors are the fault of the authors.

Achieving teaching excellence takes time. It requires honest reflection, careful listening to students, and repeated revision and fine-tuning of approaches. But, what a worthy goal to pursue!

DEDICATION

To Kelle, who is the most important teacher in Maxwell Alexander's worldly classroom.

Tim Slater
University of Arizona

To my wife, Monique, for her love and support, and to our children, Lara, Nickolas, and Joshua, for continually reminding me of the importance of cultivating curiosity in all learners.

Jeff Adams
Montana State University

Chapter 1
Introduction

That initial moment when you first capture your class's attention and welcome them to ASTRO 101—an introductory survey of astronomy for non-science majors—is unique; it is the only moment all semester that every student in your class understands absolutely everything that you so far wished them to learn! After that initial welcome, most of us have experienced a growing sense that too many of our students "just don't get it," a sense that is often painfully confirmed on the first exam. This is not to say that we haven't all taught marvelous students who were thoroughly engaged by our teaching and who repeatedly demonstrated their understanding and enthusiasm in class and on examinations. However, for must of us, this is not the norm and we spend a good deal of time worrying about what we can do to help our students achieve levels of understanding that we deem appropriate. At times we even bemoan that we seem to be lecturing to a sea of unprepared, unmotivated, and generally apathetic students who fail to appreciate the enormous efforts we put into such things as creating visually stimulating lectures with the latest Hubble Space Telescope images or developing extensive World Wide Web (WWW) sites to support our course. Is there a simple answer? Of course not! Are there things we can all do to increase students' learning in our classes? We are convinced there are! However, we are equally convinced that the first step toward improving student outcomes is a refocusing away from what <u>we</u> do to what our <u>students</u> do. It is the recurring mantra of this book that

> ## It's not what the instructor does that matters; it's what the students do.

For many faculty, this represents a major paradigm shift in how teaching is viewed—a shift from *teacher-centered education* where the professors teach and the students learn, to a new perspective of *learner-centered education*. Learner-centered education is a perspective that places the students on center stage in the classroom and places the professor as a guide rather than the central dispenser of all knowledge. It requires a revised role for faculty in which much of their limited teaching time is reallocated from providing lucid explanations to engineering "mind-on" learning experiences that challenge students' preexisting beliefs and encourage them to practice reasoning and critical thinking skills within the context of astronomy.

The three overarching ideas about student learning that frame this book can be summarized as follows: (1) Students have preexisting ideas about how the world works that must be attended to for meaningful learning to occur; (2) students need to have repeated exposures to apply complex ideas in multiple contexts; and (3) students need frequent feedback from professors and each other to help them learn to monitor their own progress and recognize when they need remediation. Our goal in writing this book is to highlight examples and strategies that will help you make a shift to learner-centered teaching in ASTRO 101. We'll hopefully even manage to be humorous now and then, although our confidence is pretty low in this regard based on our prior experience!

As an initial starting place, we provide an overview of who your ASTRO 101 students are and the sorts of things they are interested in (Chapter 2). One of our biggest revelations is that we now recognize that the overwhelming majority of our nonscience major undergraduates enrolled in our introductory astronomy survey courses are quite unlike us. In particular, the histories, mysteries, and problems of astronomy that most engage astronomers are not the same issues that capture our students' interest or naturally motivate them to engage in studying astronomy. These perspectives are provided with the intent of helping you articulate how you want your students to be different as a result of taking ASTRO 101 in the form of goals and objectives.

Equipped with an understanding of who your audience is and where your ASTRO 101 course fits in the lives of your students, Chapter 3 outlines some of the relevant results of physics and astronomy education research efforts that underlie and motivate many of the ideas we share in this book. These include a list of many of the most common student misconceptions in introductory astronomy, how teaching strategies that move the students from a passive role to an active role have significant and long-term impacts on learning, and how collaborative group learning approaches can help students who have traditionally struggled in ASTRO 101 make substantial improvements in their learning.

Just as a well-thought-out business plan is the foundation of an making consistent and effective business decisions, so too can a well-designed ASTRO 101 syllabus put students in a well-placed position to succeed. In Chapter 4, we show you how you can make your life in managing a large-lecture course infinitely easier by designing a syllabus around your course goals that simultaneously simplifies dealing with students so that you have more time to focus on the learning process.

Most of us spend considerable time lecturing to our students—often regardless of class size. Chapter 5 outlines some of the most common mistakes faculty make when trying to involve students in the class and outlines learner-centered principles that you can use to adopt a interactive and engaging approach to your class.

Probably the most well-researched learner-centered instructional technique is that of collaborative group learning. Collaborative groups take advantage of the social interactions our students. From an "in the trenches" perspective, Chapter 6 describes how we use both highly structured collaborative learning activities and open-ended tasks to help students reason critically about topics in astronomy. Additionally, we include in Appendix E nearly 30 open-ended collaborative learning tasks that are classroom ready for you to use in your course right away.

Chapter 7 addresses one of our most difficult tasks when teaching ASTRO 101—the development of multiple-choice examinations. Certainly, most of us would prefer to use other approaches to testing and grading, but the large class sizes many of us encounter make multiple-choice tests a necessity. We provide a basic primer on how to create more effective items and how to analyze your test statistically to see if it is actually telling you what you want to know about your students' progress.

Going beyond the multiple-choice test, Chapter 8 describes effective alternatives to multiple-choice tests. These alternatives include assigning and grading essay questions, numerical problems, portfolio assessments, performance tasks, and concept maps. We also provide timesaving strategies for grading these types of assignments while maintaining objectivity and fairness.

Chapter 9 provides easy-to-implement techniques and tools to conduct course assessments to help you monitor the strengths and weaknesses of your learner-centered course. These include numerical and open-ended surveys you can use to learn how much students are enjoying the class and can give some insightful guidance as to which course elements students feel are most beneficial in helping them learn.

Effective teaching is beginning to catch on at even the most traditional colleges and universities. Chapter 10 provides a structure for you to demonstrate your effectiveness in teaching. We describe one possible approach to creating a teaching portfolio that helps you reflect on your growth and improvement as a teacher.

We hope that you recognize that implementing every one of this book's recommendations is an impossible goal. Rather, it is our vision that you can select a big idea or two each semester and adapt them to work in your specific teaching environment. You certainly do not need to include collaborative group learning tasks, ConcepTests, and portfolios into every class meeting—nor should you. What is most important is that you take the time needed to adequately reflect on what is working in your classroom and what needs improvement in order to move your ASTRO 101 course to a more learner-centered environment. Given this, let us begin!

Questions to Think about BEFORE You Read This Book

- **What is my greatest strength as a teacher?**

- **How do I want my students to be different as a result of taking my ASTRO 101?**

- **If I walked into the library and saw a group of my students studying astronomy, what exactly would I like to see them doing?**

- **What would I most like to change about my ASTRO 101 course?**

Chapter 2
Goals and Objectives

What is the single most important thing you can do to improve your ASTRO 101 course? Take the time to write down the goals for your course. A well-known adage from trainers in physical fitness and advisors in financial planning is that "a goal isn't a goal unless you actually write it down—otherwise, it's just a wish." This idea applies equally well to teaching ASTRO 101 because if you don't know where you want your students to get to, how will you know if they made it?

The goals faculty set for ASTRO 101 vary widely. Perhaps you want to have the most popular class on campus with the highest enrollments or perhaps you want to be known for attracting the most students from the business college to become astronomy majors. Maybe, even, you want to have the reputation for giving the hardest tests of any general education course.

> **When thinking about your course goals, it is often helpful to consider how you want your students to be different as a result of taking your class.**

Whatever you choose as your overarching course goals, we do suggest that you center your goals on students rather than on yourself. In other words, it is often helpful to consider how you want your students to be different as a result of taking your class. Do you want them to be able to point out constellations to their friends and family or do you want them to be able to explain the inferential evidence that suggests dark matter abounds in the universe? Or, maybe both are appropriate.

When we survey astronomy faculty about what goals are most important to them, we find that they fall into three broad categories (viz. Adams, Brissenden, Duncan, & Slater, 2001). The first is that students understand the big ideas in astronomy. These big ideas most often include the electromagnetic spectrum as a tool, size and scale of the cosmos, spectroscopy, and cosmology. The second common category is that students understand something about how science is done. This involves understanding and appreciating the nature of science, the scientific method as applied to astronomy, the influence of technology, weaknesses of pseudosciences, and careers in astronomy. A third goals category relates to engendering positive student attitudes about astronomy—and science in general. Many faculty describe encouraging students to become life-long learners in astronomy, choosing to read news and magazine articles on astronomy, visiting museums and planetaria, and maintaining a desire to look through telescopes such as at a local astronomy club. Let us emphasize again that these overarching faculty course goals are about students and, although measuring the degree to which they are achieved might be difficult (more on this later), they do provide guidance for curricular decisions. No matter what your goals are—and we think you could do a lot worse than adopting some version of the ones just listed—it is always worth sharing them with your students and asking yourself exactly how each element of the course contributes to those goals.

WHO ARE YOUR STUDENTS?

Let's take a step back from goals for a moment because how you structure your course goals depends quite a bit on who your students are. In other words, your goals might look quite different depending on whether your course is filled with science majors at M.I.T. or filled with nonscience students at the University of Arizona.

So, who are your students? As part of a national survey of astronomy knowledge, Grace Deming and Beth Hufnagel (2001) collected student demographic information. In their wide-ranging survey, they found that ASTRO 101 student demographics closely mimic the general population of undergraduates across the country. The slight majority of ASTRO 101 students are women (52%), more than 65% of students are under 20 years of age, and 92% report that ASTRO 101 is their first college astronomy course. The distribution of ASTRO 101 student majors also reflects national norms, where 85% of undergraduates having declared majors outside of science, engineering, or architecture. Thirty-five percent of students are majoring in the humanities, social sciences, and the arts. Few students have confidence in their abilities in mathematics and science. In fact, whereas only 41% of ASTRO 101 students rate themselves as "good" or "very good" at mathematics, even fewer (34%) view themselves as such in science. In short, what these students are interested in and how they view themselves performing in science courses is far from optimal— indeed, these students present an arduous but worthy challenge to the ASTRO 101 teacher!

We would be somewhat surprised if the students at your institution were far different than this. However, we encourage you, at a minimum, to talk to colleagues who have taught the course before, talk to department administrative assistants who handle student paperwork, and survey your course enrollment lists to look at the distribution of declared student majors. You will find that understanding who your students are and where they are coming from will bring you much closer to meeting your course goals because you can help them make connections between their academic majors and the realm of astronomy. A few examples of making the course relevant to nonscience majors are worth noting. With some effort one can relate the interests of photography and film majors to astronomy through art and science fiction (an archive of digitized space move clips is available online at URL: http://graffiti.cribx1.u-bordeaux.fr/roussel/anim-enf.html). Similarly, recent news reports about the space telescope or discoveries from an interplanetary probe can be highlighted for journalism majors. Even business and accounting majors can be enticed by taking a moment to point out how stock prices for major contractors such as TRW and Lockheed Martin change when scientific or technological advances are made.

The one thing that we are reasonably confident in asserting is that few of your students are future physics and astronomy majors. This is important because you need to appreciate just how different you are (or ever were!) from most of your students. Their interests and talents are different from yours and most of them don't learn science the way you were able to. Although we firmly believe in the power of reflection as a means of directing change, it can actually be counterproductive to think back to when you were a student and try to emulate the teaching style and seemingly lucid explanations of those professors who most affected you. Just because it worked for you doesn't mean it will work for your students and, even worse, it may mean that it won't work for your students at all! To teach your students, you need to find out what works for them.

WHAT ARE YOUR STUDENTS EXPECTING TO LEARN?

Probably the most common reason that faculty receive low course evaluations from students is an enormous mismatch in what faculty and students expect ASTRO 101 to be about.

Box 2.1 Student Expectations

The Most Frequent Things Students Expect to Learn in ASTRO 101	
▪ constellations	▪ Moon
▪ stars	▪ Sun
▪ planets	▪ weather
▪ galaxies	▪ atmosphere
▪ black holes	▪ UFOs & the unexplained
▪ comets	

NOTE: Midcourse surveys reveal that students thought they would learn more constellations and that they are surprised that the other topics were covered in such depth.
Adapted from Lacey and Slater (1999)

We strongly urge you to conduct a brief written survey of your students asking them what they would most like to learn in ASTRO 101. In truth, you don't have to change any of your course plans fundamentally; however, you can point out when you get to certain topics that this is what they stated was a high priority. A truism exists among many college faculty that so-called great teachers are really just marketing and sales experts. If there is indeed any truth to this, the lesson is that you will benefit greatly from figuring out how to create a win-win situation for you and your students by matching students' perceptions with the structure of your course. Even in cases where you will not be addressing a particular topic of interest to your students, it helps to know this in advance so that you can acknowledge it to your class and explain why you have made this choice.

As a last suggestion, if your course fulfills a general science requirement as part of the overall college or university goals, other faculty have found it surprisingly useful to look at the overarching college or university goals for general education classes. Be sure that your goals are aligned with those written elsewhere. You might even find just the exact language you need to focus your own course goals.

COMPOSING LEARNING OBJECTIVES

Once you have written down your three, four, or five overarching course goals, the next step is to decide the specific learning objectives for your students—generally related much more to content—that will contribute toward their reaching the course goals. The writing of learning objectives should not be a tedious or time-consuming task and yet it is one that can prove enormously valuable. Learning objectives should be written as a guide for you and your students about what *specific* aspects of the course are important. And when we say specific, we mean specific. *Understand Kepler's laws* is far too general. *State Kepler's first law, use Kepler's second law to reason about the position and motion of orbiting bodies,* and *apply Kepler's third law to the motions of asteroids* are learning objectives that clearly tell the student what they should be studying and at what level of proficiency they need to reach.

Writing learning objectives that simultaneously guide how you present material, help students monitor their own learning, and inform your testing strategies is somewhat of an art form that improves with practice (and, dare we say, perhaps some friendly peer review). One way that faculty specify the levels of complexity and depth students need to achieve is to employ *Bloom's Taxonomy of Educational Goals and Objectives.* Bloom (1956) defined "understanding" at six hierarchical levels: knowledge, comprehension, application, analysis, synthesis, and evaluation. The first three levels, sometimes called the lower order thinking levels, emphasize recall, literal translation, and application of concepts to well defined situations. The second three levels, often referred to as the higher order thinking levels, focus on breaking apart complex ideas for use in novel situations, and integrating ideas across numerous concepts. Although most of us would agree that ASTRO 101 should focus on the higher-order levels, you will likely find that writing learning objectives at the lower-order levels is many times easier than writing objectives at the higher-order levels. More important, you will also find that it is infinitely easier to create test items that probe the lower-order levels than the higher-order levels. As many of our nonscience majors enter our courses with almost no basic knowledge of astronomy and very naïve views of science, it is essential that students gain some lower-order knowledge—and that you assess this knowledge. However, this does not mean that your course should be limited to lower-order knowing only; selecting specific topics in which to expect students to gain higher-order understanding enriches your course and, many would argue, is what makes it an engaging college experience.

Box 2.2 Bloom's Taxonomy of Educational Objectives for Knowledge-Based Goals for "Understanding the Seasons"

Level of Expertise	Description of Level	Example of Measurable Student Outcome
Knowledge	Recall or recognition of terms, ideas, procedure, theories, etc.	Student states when the first day of spring is.
Comprehension	Translate, interpret, extrapolate, but not see full implications or transfer to other situations, closer to literal translation.	Student describes what the summer solstice means.
Application	Apply abstractions, general principles, or methods to specific concrete situations.	Student explains why seasons are reversed in the southern hemisphere

Analysis	Separation of a complex idea into its constituent parts and an understanding of organization and relationship between the parts. Includes realizing the distinction between hypothesis and fact as well as between relevant and extraneous variables.	Student can analyze what Earth's seasons would be like if Earth's orbit were perfectly circular.
Synthesis	Creative, mental construction of ideas and concepts from multiple sources to form complex ideas into a new, integrated, and meaningful pattern subject to given constraints.	Given a description of a planet's seasons, student can propose plausible orbital and tilt characteristics.
Evaluation	To make judgment of ideas or methods using external evidence or self-selected criteria substantiated by observations or informed rationalizations.	Student can distinguish what would be the important, and irrelevant, variables for predicting seasons on a newly discovered planet.

Adapted from Brissenden, Slater, and Matheiu (2002).

We have included one of our lists of ASTRO 101 learning objectives in Appendix F. If someone were to adopt these learning objectives uncritically for his or her course, then we would have failed in our message. Your learning goals should guide your course sequence and inform your students about what is important. Similarly, well-articulated course goals and objectives will guide you when writing exams. As a matter of apology, we should mention that our style of writing learning objectives does not correlate highly with the common wisdom for composing each learning objective such that it is measurable (e.g., The student will identify the end states of stars with masses $0.5M_\odot$, $1M_\odot$, $8M_\odot$, and $20M_\odot$). Moreover, when we have shared our list of course objectives with colleagues, we've had responses ranging from a shocked, "you *only* cover 24 concepts in your course" to an equally shocked, "how could you *possibly* cover 24 concepts in your course." Along these same lines, we've had students attempt to memorize the course objectives verbatim, not realizing that the words themselves are not the content of our course. Despite these issues, we provide these to give you a feel for what a list of learning objectives might look like.

How Much Math Are You Going to Include?

One of the outstanding questions—and it is directly related to how you write your goals and objectives for ASTRO 101—is what sort of mathematics to include in your course. We described earlier that the vast majority of students who take ASTRO 101 have limited backgrounds and woefully low confidence levels in mathematics. It probably comes as little surprise that the appropriate mathematical level at which to teach introductory astronomy is a topic hotly debated among astronomers.[1] At the risk of over simplifying the positions, one side suggests that teaching astronomy without mathematics is merely pandering to students and that asking them to complete calculations will enhance their mathematical abilities and appreciation of the science. The perspective is that astronomy is a quantitative science expressed most elegantly in the language of mathematics, and students should be exposed to this beauty. The contra view holds that college students have already had more than a decade of mathematics instruction and that a heavy emphasis on mathematics in the astronomy survey course will only serve to convert them from students who think they "cannot do math" into students who are completely convinced of that fact. This perspective holds that college courses for nonscience majors should avoid off-putting mathematical rigor in favor of a descriptive approach that is also designed to improve students' attitudes toward astronomy, and science in general. Of course, most astronomy instructors recognize that there is some validity to both perspectives and therefore attempt to strike a tricky balance between a course with too much math, which confuses and scares off the students, and one with too little math, which fails to reflect the true nature of the discipline and may well sell many of the students short.

We would like to suggest an alternative perspective—one that first distinguishes between *mathematics* and *arithmetic*. When astronomers ask if we use "math" in ASTRO 101, we are never really sure how to answer. To set up a dichotomy, let us first define *arithmetic* as the process of performing an algorithm to generate a numerical result, a process our students call "plug and chug." A standard example is requiring students to

---

[1] Much of the material in this section has been adapted from the article "Mathematical Reasoning Over Arithmetic in Introductory Astronomy," The Physics Teacher, **40**(5), 268, 2002, T. F. Slater and J. P. Adams. Reprinted with

calculate the length of the semimajor axis for an asteroid given its orbital period. Other examples include asking students to compute the force of gravitational attraction between Pluto and Charon, the apparent magnitude of a star at a given distance and absolute magnitude, the wavelength of a photon with a particular frequency, and a star's luminosity given its temperature and radius.

Although mathematically sophisticated students can "play" with the required formulae and actually seek meaning in the algebraic symbols, this is most likely not the case for the numerous math-phobic students we frequently find in our courses for nonmajors. With their calculators in hand, and a fair degree of coaching, students can indeed learn to perform the operations needed to get an acceptable numerical result. However, we submit that successfully performing algorithms does little to enhance students' understanding of the underlying concepts; we do not include such calculations in our courses.

In contrast to this computational-calculator view, if we define *mathematics* as the study of patterns and a language used to communicate ideas, then we definitely include considerable mathematics in our introductory astronomy course. One of our goals is for students to be able to articulate relationships between variables to reason about the relationships between physical variables. We have found that, with some effort, we can ask students to do this without relying on calculators or formulas. For example, rather than giving students the formula for the Stefan-Boltzmann law for relating luminosity to temperature and surface area, we ask students to reason about the relative diameters of stars at the same temperature when given their comparative luminosities. Other examples of the kinds of questions we focus on, as well as the kind of reasoning we would expect, are listed in Box 2.3.

It is difficult to underestimate how intensely the majority of ASTRO 101 students lack confidence with concepts in mathematics. To demonstrate one way we introduce students to mathematical reasoning without arithmetic is to pose the following question: "What would a graph of astronomers' weights versus heights look like compared to a graph of astronomers' IQs versus heights?" We ask them to label the appropriate axes on the graphs shown in Box 2.4. When first presented with such a task, it is common for students to throw up their hands in despair but, working in collaborative teams, most are able to reach an understanding on the nature of correlated versus uncorrelated data—an understanding that it is all too easy to assume students will already have.

Goals and Objectives 13

Box 2.3 Example Questions Emphasizing Mathematical Reasoning

Question Prompts	Expected Reasoning
Sirius has an apparent magnitude of about –1.5 and an absolute magnitude of about +1.4. Would you have to wait more or less than 30 years for a radio signal to arrive from Sirius?	ANS: As absolute magnitude is defined as the apparent magnitude at 32.6 ly (light-years) and because the apparent magnitude is brighter than the absolute magnitude, the star must be closer than 32.6 ly so the radio signal would likely take less than 30 years.
You observe two stars with the same absolute magnitude and determine that one is a type B star while the other is a type G star. What can you conclude about the sizes of the stars?	ANS: The type G star has a lower surface temperature than the type G star. The only way it can emit the same amount of light energy it to be larger than the type B star.
You use the method of spectroscopic parallax to determine the distance to an F2 star as 43 parsecs. You later discover that the star has been misclassified and is actually a type G7. Is the actual distance to the star greater than or less than 43 parsecs?	ANS: Because the star is on the main sequence, a lower temperature also means a lower luminosity. When the star was thought to be more luminous, its distance had to be 43 parsecs to explain its brightness as observed from Earth. As it is actually less luminous than was thought, it must be located closer to explain its appearance.
Hadar, at a distance of 90 parsecs, has absolute magnitude is –4.1. Which of the following is most likely Hadar's apparent magnitude? a. –8.8 b. –3.9 c. 0.6 d. 85.9	ANS: If it were 10 parsecs away, it would appear magnitude –4.1. It is nine times farther away than that, which means we should receive 81 times less light. 100 times less light would be a 5-magnitude difference so Hadar's apparent magnitude should be a little less than 5 magnitudes greater. (c) is most reasonable.
Canopus has apparent magnitude –0.7 and absolute magnitude –3.1. Is Canopus located more or less than 10 parsecs away?	ANS: Canopus would appear brighter at a distance of 10 parsecs than it appears at its true distance. It must be located more than 10 parsecs away.
Ross 128 has a parallax angle of 0.30 arcseconds as measured from Earth. If an observer on Mars could repeat the measurement, would the parallax angle be greater or less than 0.30 parsecs?	ANS: As the observer would be moving a greater distance between observations, the star would appear to move more compared to the background stars. Its parallax angle measured from Mars would be greater than 0.30 parsecs.
You observe two Cepheid variable stars, A and B, which appear the same average brightness. Star A has a bright-dim-bright period of 5 days, while Star B has a bright-dim-bright period of 18 days. Which star is located closer to Earth?	ANS: From the period-luminosity relationship for Cepheid variables, we know that star B is more luminous than star A. As star B appears the same brightness, it must be located farther away.

Reprinted with permission from the 'Physics Teacher' 40(5). Copyright 2002. American Association of Physics Teachers.

It is our contention that not only can conceptual questions require mathematical reasoning but that this reasoning is often more sophisticated than what is required to produce answers for more traditional quantitative problems requiring the use of formulas. Helping students think in terms of mathematical patterns and relationships is not easy, but it seems to be many times more rewarding that meaningless calculations. Similarly, questions specifically focusing on reasoning do in fact mirror the discipline of astronomy much more appropriately than do simple computations.

Box 2.4 Mathematical Reasoning without Arithmetic

"One of the following is a graph of astronomers' weights versus heights while the other is a graph of astronomers' IQs versus heights. Identify the graphs, label the axes, and explain your reasoning."

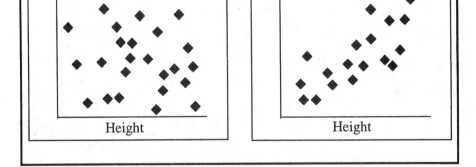

Using a philosophy of *mathematics over arithmetic* has important consequences for writing your ASTRO 101 learning objectives. For example, a recurring coffee-room debate is whether students should learn luminosities and flux or absolute and apparent magnitudes. When employing *mathematics over arithmetic* perspective, there is no debate. Asking students to calculate the luminosity of a 3500K star with a diameter of 10^7 km is an exercise in arithmetic. However, asking students to estimate the distance to a star that has an apparent magnitude of 4 and an absolute magnitude of 5 is perfectly appropriate (*the answer we are looking for is "a little less than 10 pc"*). Other places in our course that we infuse mathematics include historical measurements about the relative distances to Sun and Moon, classifying morphological characteristics in galaxies and spectra, and comparing intensity versus wavelength curves for various objects—all without a single formula. In short, if the students can successfully and powerfully use the concept, then we include it.

INCLUSIVE ASTRONOMY TEACHING

We would be negligent if we avoided mentioning that there is a longstanding tradition of teaching astronomy from a Eurocentric perspective. In and of itself, adopting a European emphasis to ASTRO 101 is not a bad thing. However, teaching from this perspective sometimes makes it difficult for the diverse student body in ASTRO 101 to see connections between astronomy and their personal heritage. As a discipline, astronomy has grown and benefited from a wide range of approaches, and our courses need to reflect this in some way.

Box 2.5 Inclusive Astronomy Teaching

Some Questions to Ask Yourself about Being Inclusive
- Do I use *he* and *she* equally in my examples?
- Do I use ethnically diverse names for people in my examples?
- Do I call on men and women equally in class?
- Do I mostly describe the work of male astronomers?
- Do I show students that science is a creative endeavor by many people?
- Do I describe telescopes operated by countries other than the United States?
- Do I point out the historical developments from cultures other than European-American?
- Do I provide multiple ways for students to obtain the concepts in addition to attending lecture?
- Have I made specific arrangements for physically disabled students to look through telescopes?

From a perspective of teaching that focuses on students, inclusive teaching that recognizes a diverse student body is paramount. Most faculty are aware that they should vary equally between the pronouns *he* and *she* and that a wide range of ethnic names should be used in examples. They are also aware and that the historical contributions of underrepresented groups and non-Western cultures should be included in courses (e.g, female astronomers' contributions to spectral classification, Mayan time keeping, Polynesian voyaging). However, work by Bianchini and her colleagues (2002) suggests that the primary reason that faculty do not emphasize diverse perspectives on science is that faculty are not able to find time to educate themselves on these issues, and aggregate work desperately needs to be done in this regard for interested faculty.

Moreover, the road to providing a truly inclusive ASTRO 101 class is a much longer one than most of us acknowledge. Authors, such as

Bianchini et al. (2002), Rosser (1997), and Banks (1999), suggest that merely adding side notes about the work of underrepresented people has merit, but is insufficient. They argue that teaching inclusively should focus on enabling students to consider concepts from diverse perspectives and to appreciate that knowledge is socially constructed. As much as we would like to, we are not in a position to provide guidelines on how to make ASTRO 101 more inclusive. All we can hope to do at this point is to make you aware that your class will be greatly improved if you can find ways to include astronomy as a diverse endeavor in which people of many different backgrounds engage.

Box 2.6 Principles of Good Practice

The *Seven Principles for Good Practice in Undergraduate Education*

1. Encourage student-faculty contact.
2. Encourage cooperation among students.
3. Encourage active learning.
4. Give prompt feedback.
5. Emphasize time on task.
6. Communicate high expectations.
7. Respect diverse talents and ways of learning.

Adapted from Arthur W. Chickering and Zelda F. Gamson (1987).

Chapter 3
Teaching for Understanding: Recent Results from Physics and Astronomy Education Research

Over the last two decades, our scientific community has witnessed an explosive growth in the number of scientists who are adopting research in teaching and learning as their principal area of academic scholarship. In particular, recent national conferences of the American Association of Physics Teachers (AAPT) have seen physics education research (PER)[1] presentations and PER participation go from being barely visible to dominating many conference attendees' schedules. The AAPT, with leadership from the PER community, is even publishing *Physics Education Research—A Supplement to the American Journal of Physics* to serve this community. Some of the recent results resulting from this flurry of activity have significant implications for teaching ASTRO 101; we summarize some of the most influential ones to provide the reader with a context for the recommendations in the following chapters.

Students Can Successfully Solve Seemingly Complicated Problems With No Meaningful Understanding

Although certainly not the first to present these ideas, probably the most publicized introduction to the impact of research in PER is the story of awakening told by Harvard physics professor Eric Mazur (1996). Mazur, a respected research physicist and award-winning teacher, had always enjoyed teaching introductory physics courses, found his students could

---

[1] Many members of the *Astronomy Education Research* (AER) community identify themselves as part of the *Physics Education Research* (PER) community while others have called for a new designation of a combined *Physics and Astronomy Education Research* (PAER) community. For the present purposes, we use PER to include astronomy and space sciences.

solve complicated physics problems on his tests, and consistently earned
high marks on his end-of-term course evaluations from students. In short,
by every traditional measure, he was an excellent teacher and had every
reason to be pleased. His view of his own effectiveness was, however,
about to change.

In the 1980s and early 1990s, David Hestenes and his colleagues at
Arizona State University developed several conceptual tests—the results of
which suggested that students could earn high marks in physics courses yet
retain only a superficial understanding of physics. This work culminated in
the creation of the *Force Concept Inventory* (FCI), which has become a
standard tool in PER for assessing conceptual learning in the domain of
basic mechanics (i.e., motion and Newton's laws). The great attention
afforded this 29-item multiple choice test derives from four factors: (i) there
is widespread agreement within the scientific community as to the
importance of the content being assessed; (ii) the test items are deceptively
simple leading most instructors to greatly overestimate the likely success
rate of their students; (iii) the results are highly consistent across a large
constituency ranging from classes in high schools to large research
institutions; and (iv) students' responses are highly resistant to traditional
instruction. Although the students' low scores on the FCI were troubling,
what was more troubling was the relatively small improvement that
occurred as a result of instruction—a result demonstrated not only at
Hestenes's own institution but also for students at Harvard.

As a skeptical scientist, Mazur, upon hearing of Hestenes's results,
developed a simple test in electric circuits to test this hypothesis of low
gains after lecture for his students. The test required students first to
perform numerical calculations for a complicated electric circuit and then to
make general statements about the brightness of light bulbs in a very simple
circuit. To Mazur's great surprise, his Harvard students could easily solve
the numerical problem but fell woefully short for the simple, qualitative
problem. In one case, a student asked Mazur if she should answer the
qualitative question in the way that he taught her or in the way that she
personally thought about electricity. In a process of self-discovery that
continues to repeat itself again and again to scientists all over the world,
Mazur realized that clearly showing students how to solve numerical
physics problems was insufficient for developing their deep conceptual
understanding of fundamental physical concepts. His students' deeply held
misconceptions were easily masked by their facility to algebraically
manipulate equations.

In reflecting on the results of these and many other similar
experiments, a simple answer has repeatedly emerged—students fail to

develop a deep understanding because most traditional instructional approaches do not actually require them to do so! Students are in fact able to succeed on most faculty-created tests by memorizing a short list of facts and by being able to recognize homework-like problems and mimic solution algorithms. For students, the good grades that result from this strategy— one that is in most cases operates at a subconscious level—reinforces their belief that they are learning successfully. In fact, questions that probe (and generally uncover) deeply held misconceptions are often viewed by students as "trick questions." Moreover, for faculty, their belief that student learning is occurring is shored up in the same way. Students produce solutions to complex problems and, because the reasoning strategies that students use to arrive at those solutions (often a matching process to similar problems) are not evident, faculty reasonably conclude that students have an understanding of the underlying physical concepts and are reasoning the solution in the same way that faculty do.

Students Have Numerous Preexisting and Inaccurate Beliefs and Understandings that Interfere with Learning

A cornerstone to a constructivist approach to instruction is to identify and confront students' initial ideas about the world. In years past, faculty have looked at students as *tabla rasa*—the idea that students enter the ASTRO 101 as "blank slates" on which knowledge can simply be written (Mestre, 1991; Mestre & Touger, 1989). Although certainly not the first nor the most comprehensive study, the most public demonstration of how initial ideas can interfere with instruction was presented in the video "Private Universe" produced by Philip Sadler and his colleagues at Harvard. In this video, both Harvard college graduates and middle school students are asked to describe the reason for the seasons and the cause of Moon phases and they perform, to most viewers' great surprise, very poorly. This provocative video then goes on to show how instruction that does not adequately consider initial student ideas is unsuccessful in altering those ideas.

As the culmination of more than ten years of work, Neil Comins (2001) has identified more than 1600 misconceptions students have about astronomy. As if the presence of so many scientifically inaccurate ideas were not troubling enough for astronomy faculty, Hufnagel (2000) and her colleagues studied students at a wide variety of colleges and universities around the country over a wide range of introductory astronomy topics using a widely available survey called the *Astronomy Diagnostics Test*

$(ADT)^2$. They report what most of us have always feared—that most astronomy lecture courses have only a limited impact on students' basic astronomy knowledge beyond the accumulation of facts.

Box 3.1 Thirty-three Common Misconceptions

1. Seasons depend on the distance between the Earth & Sun.
2. There are 12 zodiac constellations.
3. The constellations are only the stars making the patterns.
4. The North Star is the brightest star in the night sky.
5. Stars last forever.
6. All stars are same color.
7. Stars really twinkle.
8. All stars are isolated.
9. Pulsars are pulsating stars.
10. Asteroid belt is densely packed, as in Star Wars.
11. Meteors, meteorites, meteoroids, asteroids, and comets are the same things.
12. A shooting star is actually a star falling through the sky.
13. Comet tails are always behind the comet.
14. Comets are burning and giving off gas as their tails.
15. All planetary orbits are circular.
16. All planets have prograde rotation.
17. All moons are spherical.
18. We see all sides of the Moon.
19. Ours is the only moon.
20. Spring tide only occurs in the spring.
21. Only the Moon causes tides/the Moon has no effect on tides.
22. High tide is only between the Earth and Moon.
23. Once the ozone is gone, it's gone forever.
24. Mercury is hot everywhere on its surface.
25. Giant planets have solid surfaces.
26. Saturn is the only planet with rings.
27. Saturn's rings are solid.
28. Pluto is always the farthest planet from the Sun.
29. The Sun primarily emits yellow light.
30. The Sun is solid and shines by burning gas or from molten lava.
31. The Sun always rises directly in the east.
32. Black holes are empty space.
33. Black holes are huge vacuum cleaners in space, sucking everything in.

Adapted from Neil Comins's book *Heavenly Errors*, Columbia University Press, 2001.

We now understand that if students are going to abandon their self-formed and scientifically inaccurate ideas, they must not only be dissatisfied by the current knowledge, but the new knowledge must fit with other existing ideas and be more productive in terms of understanding a

---

[2] The *Astronomy Diagnostics Test* (ADT) is included in Appendix C at the end of this book and can be downloaded from http://solar.physics.montana.edu/aae/adt/.

wide range of phenomena (Posner, Strike, Hewson, & Gertzog, 1982). Otherwise, new ideas presented in class will not be successfully integrated into existing knowledge, will not be easily accessed when required, and even if memorized for the exam will be quickly abandoned afterward. This applies no matter how clearly the lecturer describes the concept. This recognition has led faculty and PER researchers to develop teaching strategies, often known as "cognitive change" strategies, that *elicit* student ideas, *confront* student ideas to help students see inconsistencies, and guide students to build accurate ideas that *resolve* these inconsistencies. Although not the only pedagogical approach to address issues of cognitive change, this is the approach that has been used most widely in recent reforms of physics and astronomy curriculum and is one that can lead to demonstrably better student learning.

Most Students are Dissatisfied with Our Introductory Science Courses

If you are not concerned yet, hang on. The situation is actually worse than it seems. Not only are students leaving our courses with a far lower level of understanding that we would like, but work done independently by Shelia Tobias (1994) and Reddish, Saul, & Steinberg (1998) suggests that too many students leave introductory science courses with more negative attitudes toward science than they originally arrived with. Tobias reports that students find that the path to an A in most introductory science courses involves memorizing what faculty write on the board and then regurgitating it back on a test—a process that most students find dreadfully boring. Reddish reports that students enter physics courses excited to learn about how mathematics and science are integrated and how science describes phenomena in the "real world." To his great dismay, he found that students leave physics courses believing that success in physics is about "finding the right formula" and that contrived physics problems have little to do with the

> **In an English literature class, students are expected to read the Shakespeare, and the professor's job is to help students interpret it. In stark contrast, it appears that most faculty teaching introductory science courses believe their role is to tell students the same things that are already in the textbook.**

"real world." Perhaps surprisingly, these results apply to high-ability students as well, who often leave science majors to study in the humanities or social sciences, a result presented by Elaine Seymour and Nancy M. Hewitt in their groundbreaking book *Talking About Leaving: Why Undergraduates Leave the Sciences* (1997).

In terms of introductory astronomy specifically, students' expectations of what the course is going to be about are often quite different from what professors plan. Students most often think that learning astronomy is going to be first and foremost about learning constellations and, second, about black holes, space travel, and the Big Bang. Imagine students' surprise, and subsequent disappointment, when black holes and the Big Bang Theory usually get at most a single a lecture each and that constellations and space travel might only receive a cursory mention as a side note in lecture (Adams, Brissenden, Duncan, & Slater, 2001). Although many faculty state that one of the goals of the introductory astronomy course should be to engender positive attitudes toward science, it should be of little surprise that too many astronomy students find astronomy to be boring if the topics that are covered are based on what astronomers find interesting and not what students find interesting.

Active Engagement Strategies Work

Most calls for reform in college classes call for professors to develop courses in which students to become more active participants in their own learning. With the development of powerful conceptual tests, such as the aforementioned FCI and the ADT, it should be of no surprise that researchers have begun to compare various approaches to instruction. Time and time again, students in classes featuring some form of active learning experience perform better—using a variety of measures—and enjoy their learning more than students who have only been lectured to.

The most comprehensive of these comparisons was compiled by Richard Hake (1998) who compared 6542 students in 62 introductory physics courses across the nation. He found a significant difference between the gains demonstrated by students in classes characterized by some kind of "active engagement" compared to students in "traditional" courses as measured by the FCI. Students in "active engagement" classes had much larger pretest/posttest FCI gains than students who took physics in more traditional contexts, and the results were independent of the size or prestige of the institution.

Active Learning Lasts

A common complaint among faculty—and one that is confirmed by talking to students—is that, all too often, material is "learned" for an exam and then promptly discarded—the "cram and flush" approach. The primary goal of education is to do more than train students to take a near-term test; the goal is to change students' worldviews permanently and profoundly. In fact, an early criticism leveled against so-called active learning classes was that they were doing little more than a more effective version of cram and flush. However, there is a now general sense among the faculty who significantly modified their courses and are regularly achieving high FCI gains that they were witnessing real and long-lasting conceptual change. To assess this, one of us (JA) and colleagues (Adams, Francis, & Noonan, 1998) conducted a long-term follow-up of students' retention. If students' worldviews had not truly been changed, it was expected that they would gradually forget the "right answers" leading to large declines in FCI scores measured several years later. Remarkably, though, what this study found was that up to three years after the course was over, students retained more than 80% of their initial gains and were still scoring significantly higher than students in traditional classes immediately following instruction. In other words, students who learned via active engagement really do learn, and this learning is not superficial as many critics initially charged; it is the kind of deep learning for which we all strive.

Collaborative Group Activities Remove Gender and Ethnic Differences

One form of interactive engagement teaching that is getting a lot of attention is the use of collaborative groups. The theoretical underpinning of collaborative group learning is one that many scientists find difficult to swallow— that most students learn best through social interactions. This is antithetical to how many

Excellent Resources
Summaries of astronomy education research:
Astronomy Education Review
http://aer.noao.edu
AER Online Bibliography
http://www.cdes-astro.com
Astronomical Society of the Pacific
http://www.astrosociety.org/education/resources/educ_bib.html

physical scientists conceive of the learning process—an individual view born out of personal experience that fails to recognize just how different most us Ph.D.s in science are from the general student population (dare we say "geeks?") Some faculty are taking advantage of students' natural inclination to social learning by inserting collaborative group learning tasks right into the lecture portion of astronomy courses (Chapter 6 is devoted to this idea). From a PER perspective, Michael Zeilik and his colleagues at the University of New Mexico (1997), who had a high percentage of students from non-Caucasian ethnic backgrounds, were looking carefully at the effectiveness of using collaborative groups and concept mapping techniques in his introductory astronomy course. Results on their precourse surveys confirmed a well-known problem that females of all backgrounds and students from non-Caucasian ethnic minorities enter courses with lower preparation and more negative attitudes than Caucasian males. Serendipitously, what they were surprised to find was that, as measured on postcourse surveys, these initial differences among the populations were erased even though there was no intentional focus on these traditionally disadvantaged students.

Teaching for Understanding

The current state of affairs for most students is that *professors talk and students learn*. The studies briefly described here, and the scores of others with similar results that we didn't mention, strongly suggest that such a view of college science courses is insufficient. What we are advocating in this book is a change in faculty perspective on teaching—from a view of the professor as a *teacher-centered* dispenser of knowledge to a *learner-centered* orientation where the professor's role is to engineer productive learning environments. In short, learner-centered teaching considers the preexisting ideas students bring to class, helps students develop meaningful understanding that is flexible and long lasting through a variety of experiences, and provides a means for frequent student feedback to help students monitor their own progress. We also fully appreciate that few well-meaning professors exclusively embrace a teacher-centered view—most would agree with the fundamental precepts of learner-centered teaching. What they struggle with are strategies for translating this view into effective approaches in the large-enrollment class, where the physical environment, sheer size of the class, and even students' very traditional views of the teaching/learning process conspire to make this a daunting task.

Box 3.3 Levels of Understanding

What does it mean to understand an idea in science?
1. name
2. recognize
3. describe
4. compare
5. apply
6. generalize
7. integrate with other ideas

Chapter 4
Designing an Effective Syllabus

Fundamentally, a syllabus communicates to your students the key elements of your course[1]. It explains what they will be learning, why, and how their learning will be assessed. As such, it serves as a contract defining both what they can expect of you and what you expect of them. Although a minimalist syllabus might only list office hours, exam times, and how a final grade is calculated, we believe that the up-front time invested in creating a comprehensive syllabus that conveys to students a complete picture of your course will, in the end, pay many dividends over time. Viewed in this way, creating your syllabus is really a process of comprehensive instructional planning designed to maximize students' possibilities for achieving your stated learning objectives (see Chapter 2). This chapter is therefore really about course planning.

PRE-PLANNING

Pre-planning is the stage in which you develop (or refine if you have taught the course before) an overall picture of how you would like the course to operate and what experiences you hope to provide for your students. If you have taught the course before, this is a time to reflect on what you learned

> **Pre-planning is time spent thinking about how exactly you want your students to be different as a result of your course.**

last semester (see Chapter 9) and contemplate what new instructional strategies you might want to try (see Chapter 5). If this is the first time you have taught this course, especially if you are new to the institution (which frequently happens because ASTRO 101 is often viewed as an undesirable

---

[1] The syllabus is also important for students transferring to other institutions who need to have transfer credits assessed.

teaching assignment foisted off on new faculty) there is a great deal that you can do in terms of preliminary factfinding that will increase your chances for a successful first semester.[2]

Talk to Previous Instructors and Students

The concise catalog description is often the last place that students go to learn what to expect from your course; in reality, it is the informal student-to-student advising network where much of the real information is transmitted about courses. Therefore, you should learn all you can both about how the course has been run in the recent past and, just as important, you should endeavor to find out what the students are saying about your course. You do not want to find out two weeks into the course that the class GPA the last semester it was taught was 3.6 and that students often scheduled other classes at the same time as the lectures because the course culture was "All you need to do to get an A is to read the book the night before the exam." Although this is not an ideal situation in which to begin, it is better to know of this reputation up front so that you can deal with it. If you are faced with this situation—and assuming you think it should change—there is no simple answer. If you do decide to make a drastic change in the course's expectations right away, it will lead to tremendous student dissatisfaction and will show up negatively on your end of course evaluations. Such a situation makes it imperative that you talk over any radical changes in advance with your department head, and perhaps even your dean, to ensure that you will get the support you need if students complain.

A more general consideration that is also largely a matter of campus culture—and one that you should take seriously—is the issue of unofficial student "holidays" such as the Tuesday or Wednesday before Thanksgiving and the Friday before spring break. The degree to which you can expect students to attend class on these days depends on the institution and on the particular class you are teaching. We do not advocate specific strategies for dealing with this, which can range from canceling class to holding a major exam, but we do suggest that it is something you need to consider in developing your course plan.

Find Out Why Students Will Be Taking Your Course

Although there is a strong innate interest in astronomy among the general population, the reality is that most students take ASTRO 101

---

[2] However, if your first semester is not successful, don't get too discouraged. Even the most celebrated teachers often tell of horrible experiences when teaching a class for the first time, especially at a new institution.

because it fulfils some general graduation requirement. You might be surprised to know that, as a university general education course, there are usually additional expectations—such as writing requirements or student oral presentations—that your course must satisfy for the university. Because these requirements range widely, we strongly encourage you to investigate this at your own institution and talk to colleagues about how exactly this will affect your course design.

Beyond general education requirements, there are other elements in students' curricular requirements that could direct them into your course (e.g., additional science requirements for engineering, computer science, or pre-service education majors) and it is important to understand these before planning your course if you wish to try to make your course relevant to the students.

Find out what resources you will have

The degree of support (graders, teaching assistants, demonstration setup, audiovisual, etc.) provided to instructors in large-lecture astronomy courses varies tremendously and can have a profound effect on how you plan your course. You might wish to assign term papers but, without grading assistance, you may be forced to abandon this idea in favor of less labor-intensive assessments. Or, you might decide that student writing is so important that you approach your department head about increasing your grading support for what you see as a critical course component. One approach to dealing with this that is far too often overlooked in departments with graduate programs (and therefore a history of using graduate students as teaching assistants and graders) is to pay undergraduates to grade for you. Undergraduates are typically paid significantly less than graduate students and, by selecting your top students from previous semesters, you can often acquire high-quality help—perhaps even better than randomly assigned graduate students. It is also a good idea to devote some time wandering through the lecture demonstration room and finding out what demonstrations have been used in the past.

DEVELOPING AN INSTRUCTIONAL PLAN

As advocated earlier, the place to start is to write down your three or four most important learning goals for the course. Having these goals firmly in place will help you make decisions about what to and what not to include as you craft your instructional plan. The key elements that define your course and on which decisions must be made are as follows:

- the sequence of topics that you plan to teach (including learning objectives)
- your in-class strategies for teaching each of these topics
- the additional requirements that students will be expected to complete outside of class time (observing, homework, reading, etc.)
- the methods you will use to assess their learning for the purpose of assigning a grade

Schedule of Topics

Your syllabus should provide an outline of the topics to be covered and a schedule. At a minimum, we recommend listing the order in which the topics will be taught and, if possible, we suggest listing the weeks (or even days) on which the topics will be covered. The specificity of your schedule involves some real tradeoffs that you need to consider. Students appreciate knowing that on November 18 they will be learning about the solar cycle (and often think of a syllabus as nothing but a course schedule). However, once this level of specificity is supplied, it is important to stick to it, which really limits your flexibility to make adjustments. Generally, we suggest offering a less specific schedule (maybe even just an ordered list of topics) until you have enough experience to predict with some degree of certainty what the schedule will be. If you do not provide a detailed schedule in advance, a complete listing of dates, topics, learning objectives, and text references is easily listed on a Web site after the fact and then can serve to inform a more complete schedule next year. One word of caution is in order: If there is something special you wish to address, black holes for example, sage advice is that if you don't purposefully schedule it, it may not get done.

Box 4.1 Sample Instructional Philosophy from Syllabus

Roles and Responsibilities

In many courses you have had before, the professor's responsibility was to lecture and your responsibility was to take notes and memorize the material. Not so with this course. In this course, my responsibility is to find ways to help you learn astronomy, and your responsibility is to actively engage in your own learning of astronomy. My overarching goals for this course are for you to understand the nature of science through the eyes of astronomy; understand the big ideas and methods in astronomy; and develop a lifelong interest in astronomy and current events surrounding astronomy. To meet these three goals, I have carefully designed a sequence of mini-lectures, learning tasks, and assessment procedures as outlined in the syllabus. They include the following:

- *Active engagement with daily active learning activities.* It is my belief that you can only learn so much from a lecture, no matter how clear or entertaining. Therefore, the course is composed of a series of mini-lectures with targeted classroom activities separating each mini-lecture. These activities are designed to be completed in pairs during class by talking through the questions and writing a detailed, consensus response. Typically, you will not submit these for grading; however, the questions are quite similar to the questions you will find on the course exams and I strongly encourage you to consider these activities as a critical component to your success in the course.

- *Attendance is REQUIRED at all classes.* Because this course is built around daily activities to accompany the lecture, your attendance and participation is required at all class meetings. Each day, attendance will be taken in a variety of ways, usually through provided bubble forms. If you need to miss a class for a valid reason, contact the course teaching assistant by e-mail or phone as soon as possible to petition for excuse. Only three excused absences are allowed, and they all must be submitted prior to the class meetings missed.

- *Carefully studying the text is REQUIRED.* The course mini-lectures are designed to focus on the really difficult aspects of astronomy or to provide structure for your out-of-class study. You are accountable for all material, concepts, and interrelationships presented in the mini-lectures, the activities, and the text. Therefore, it is imperative to your success in this course that you complete the assigned readings prior to coming to class. Otherwise, the mini-lectures or activities won't make much sense. Please bring your text, along with your activities, to class each day so that you may make notes in the margins and highlight the relevant passages. It is important to remember that the exams will cover material from the text that might or might not be discussed in class.

- *Continuously monitoring of your learning with frequent (nearly daily) ungraded quizzes.* In an effort to help you monitor your learning and for me to keep a close eye on the progress of the class, you will be given ungraded quizzes covering material in the text, lecture, and tutorials. Typically, these questions will be given after a lecture and again after a tutorial, although not necessarily on the same day. These quizzes will be taken on bubble sheets so that I can record class progress. Your grades on these ungraded quizzes will not impact your course grade directly. Answering the questions as best as possible will help you understand which aspects of astronomy you understand well and which aspects need more attention in your at home studying, as well as help us tailor the course materials to be simultaneously challenging and still attainable.

In-Class Teaching Strategies

Your decisions about instructional practices will affect your syllabus in many ways. At the very least, the methods you select to teach your objectives will impact course scheduling; there is always a tension between the desire to cover the material (the entire universe in a single semester?), which is done most quickly in lecture mode, and the desire to provide students alternative learning experience to promote deep learning (see Chapters 5 and 6), which almost always require a larger allocation of time. An understanding of what topics you can reasonably expect to teach in the time you have and how various teaching strategies will affect your schedule are things that develop with experience.[3] Beyond impacts on your schedule, it is important for you to explain to your students the teaching strategies you will be using, your rationale for using them, and the resulting expectations for your students. All too often, students feel that their only job is to come to class and copy down what they must then learn—an expectation too often confirmed by experience. To move them away from this view it is important to communicate your expectations repeatedly by explaining their responsibilities and your role in helping them learn; the syllabus is a great place to begin this.

Ideally, every student would always come to class with insightful questions as a result of having already read the relevant sections from the textbook. This would allow you to spend less time describing basics and more time addressing issues of confusion or current events—something that could actually increase the overall number of topics you could plan to teach. To encourage this, we suggest you consider using regular reading quizzes on material that you have not yet covered in class. Of course, you will never get every student to be prepared for every class (or even come to every class for that matter), but you can increase the regularity and attention with which your students complete reading assignments using regular reading quizzes. These should be easily administered and graded (multiple choice is appropriate) and focus on the basic knowledge found in the chapter. The idea is to use questions that are easily answered by students who have read the relevant material carefully and to avoid both questions about isolated facts requiring memorization and those questions requiring deeper levels of understanding you'll address during class. The point of the reading quizzes is to reward students for doing their reading; if the quizzes

---

[3] Over time, we find that we have had a tendency to teach fewer topics and teach with a greater expectation of understanding. One way to accomplish this is to assign appropriate topics—primarily factual ones—as reading assignments so that class time is devoted to more substantive issues. To make this strategy successful, students must be held accountable on your tests for the reading assignments.

are so difficult that they get the feeling they would have done just as well without doing the reading then, true or not, they will be less inclined to prepare next time. For the first quiz, because it sets the expectations, it is especially important to err on the side of too easy than too hard (yes, you read that right). It is easy to overestimate your students' abilities to learn material by reading it in the book—even your best students—and therefore it is easy to set a quiz where much of the class performs little better than if they guessed.

If you make the decision to use reading quizzes, the other important decision is whether to announce them ahead of time or make them a surprise. If they are announced, they have to be regular to be effective. Making the quizzes a surprise has the virtue that there can be fewer because the possibility is always there—something of which you should regularly remind your class—but they can be perceived as unfair. If you are at a new institution and are unsure about this, we recommend asking an experienced colleague for her or his recommendation.

Out-of-Class Activities

If you have scheduled activities that take place outside of regular class time—especially ones for which students might have to arrange time off work to attend, such as observing, a planetarium visit, or group projects—these should be listed in the syllabus and repeatedly drawn to students' attention. We have found it difficult to schedule required activities in the evenings because so many of our students have legitimate conflicts such as work, childcare, and long drives. In particular, after-hours requirements preferentially have a negative impact on nontraditional students. We therefore generally make activities such as observing or a planetarium visit optional or as a package of several choices. In fact, at some institutions it is policy that no activities (even optional ones) can be scheduled outside of the times listed in the catalog.

Whether or not to assign homework is a decision that is likely to depend on your class enrollment and grading resources (including your own limited time). In the perfect world, homework should be assigned at each class meeting and collected at the next, thus providing incentive for students to engage consistently in the material within a few days of class, which greatly enhances retention. However, with large classes this is often impractical. One way to provide incentive, although not as much feedback, is to assign homework but collect it randomly based on, for instance, the chance roll of a die.

An even more resource-intensive activity, but one that many faculty consider to be worth those resources, is collecting and grading student writing. However, because of the tremendous amount of work involved in providing thoughtful and timely feedback, many other faculty avoid this completely (see Chapter 5 for an alternative in-class writing strategy). If you choose to collect writing, the single most important thing is repeatedly making your expectations clear in terms of grammar and spelling, depth of research, expected number of drafts and redrafts, amount of original material and the like. Given the chance, students, like all of us, will dash off writing assignments at the last minute. Unfortunately, though this strategy may work just fine for them in more familiar contexts such as their major, it often leads to very discouraging outcomes in science courses. Almost all faculty who have tried writing assignments in astronomy bemoan the low quality of work they receive and declare that, "Students can't write." Actually, we contend that the problem is that students don't understand science very well (they are usually nonscience majors) and that writing quality is contextual. In other words, students often don't write well for introductory science courses because they really don't know what they are trying to say. This is compounded by students' general expectations that writing is something that is important in English and history courses, not in science courses. Ensuring greater levels of success in writing assignments in your astronomy class requires careful planning (we recommend consulting someone from your institution's writing center if possible), communicating high expectations from the beginning, encouraging students to complete drafts (consider a draft workshop in which students critique each other's drafts and you collect copies of the drafts), providing high-quality student work samples, and having high grading expectations. You do not need to identify every technical error in their writing but, if a particular piece of student writing does not meet some minimum standard, you should consider returning it to be rewritten. If it does meet the minimum standard, try to avoid focusing on the small technical errors and instead read for and comment on the message the student is attempting to convey.

The most potentially time consuming, and also potentially rewarding, out-of-class activity is some form of term project, which can take a variety of forms. The most straightforward is to require a term paper that students complete based on library and Internet research. For instance, you could ask students to select a major astronomy facility or mission and report on its history, major accomplishments or discoveries, and impact on our understanding of the universe. You should require that it be typed, specify a page or word count requirement, and require a minimum of number of resources that must be cited. We also suggest that you communicate your expectations for writing quality as discussed previously.

It is worth noting that although today's students are very good at finding information on the Internet, they do not often have strong skills at synthesizing the information. One approach to helping students avoid the temptation to "click, copy, and paste" into a paper is to ask them to devote much of their time to creating arguments based on evidence. For example, because it is unlikely that there is a Web site out there that describes why the Keck is better than Hubble Space Telescope, it is reasonable to ask students to describe which telescope they would support funding, why, and what discoveries they compared to lead them to their position. Similarly, students could be tasked to create a "briefing" for their state senator on several telescopes seeking federal funding.

Instead of asking students to communicate results in a written document, many faculty find it much more rewarding for everyone involved to sponsor a poster session in which results are shared much like at a professional meeting (minus the open bar of course!). Such a format encourages creativity and the use of images, allows more naturally for group projects (thus reducing grading time), and encourages students to become more personally invested in their projects because they will be displayed publicly. We suggest that you arrange to have a large room set aside, perhaps in the student union, and invite colleagues and other physics students to view the projects and perhaps participate in the grading or even giving awards by interviewing the presenters. Consider having students submit an abstract via e-mail well ahead of the "meeting" and publish a program. The campus newspaper staff will often help with advertising and coverage.

If facilities exist, term projects need not be limited to library research but can, in fact, involve some form of student observing, which can be reported in a paper or a poster. Such research projects (see Box 4.2) must be carefully organized because many require observations over long periods of time and must be started early. Without regular reminders and, where possible, early submissions of preliminary results that you closely scrutinize, many students will allocate significant unproductive time or, worse yet, simply put it off until it is too late to do something adequate (and then come to ask you what kind of "extra credit" they can do). Using observing projects in classes with hundreds of students can be a daunting prospect. One way to introduce projects gradually is to make them optional bonus assignments. For instance, you can offer extra points (say 5 %) or allow students to replace one exam grade with an observing project. Many students may initially plan to participate but few actually follow through, which provides you a good way to try out ideas on smaller numbers of students even if you have a large class.

Box 4.2 Sample Student Observing Projects

- Sunset Azimuths Using Sketches or Photographs
- Changing Solar Altitude Using Noontime Shadows
- Local Moonrise Times versus Newspaper Reports
- Visible Magnitudes versus Photographic Exposure Times
- Delta Cepheius Light Curve
- Creating a Local Medicine Wheel
- Identifying and Grading Light Pollution Sources
- Sunspot Counts and Solar Rotation Rate
- Determining Periods of Jupiter's Moons
- Timing Bright Occultations

Determining Grades

For many students, the most important information conveyed on a syllabus is the statement defining how grades will be determined. All too often, students perceive grading as a magical process that they don't understand and one that encourages competition because grading on a curve means the failure of others is as important as personal success. We recommend making your process for determining student grades as simple and transparent as possible and routinely reviewing how the various course elements affect final grades.

If you are new to an institution, begin by reviewing your institution's grading policies, which might contain very specific requirements regarding how you must assign grades (e.g., whether or not pluses and minuses are used or if you have to record dates of failing students' last attendances). We also suggest talking to several experienced colleagues, especially those with experience teaching nonmajors, to learn something about common practices and student expectations at your institution.

The most important decision to make regarding your grading policy is whether or not to "grade on a curve" (called norm-referenced grading) or according to some predetermined scale (called criterion-referenced grading). We strongly advocate that you avoid norm-referenced grading (e.g., stating that only

> **When students ask, "What is the curve?" they are really asking how to convert their numerical grade into a letter grade.**

the top 10% of the class will receive an A) because of the loss of a sense of personal control this creates in students—their grade is now affected as much by the performance of others as by the quality of their own work. Although less well-defined, telling students that you will assign numerical grades for now and determine the letter-grade cutoffs at the end also seems arbitrary and out of the students' control. Ideally, you should contract with students on the syllabus what minimum score (percentage or total number of points) will earn them each grade and they should know how the various course elements contribute to that grade. Of course, you can always choose at the end to lower the minimum required scores if, for instance, you think that you underestimated the difficulty of the final exam. We strongly suggest you make grade modifications at the end of the term and avoid making scoring modifications to each exam except in the cases where certain items did not discriminate appropriately (see Chapter 7 for a discussion of statistical item analysis).

Box 4.4 Sample Grading Scheme

Absolute grading (no curves, no competition, it is in your best interest to help each other learn astronomy)		90%—100%	A
		80%—89.9%	B
1.Three Exams (*drop lowest*)	50%	70%—79.9%	C
2.Final Exam (*cannot drop*)	25%	60%—69.9%	D
3.Planetarium or Observatory Visit	5%	< 59.9%	E
4.Class Participation in Activities	20%		
(*attendance recorded 10 times randomly*)		No plus or minus grades	

Checking Your Grade

Your current course grade can be found on the Internet. Otherwise, course grades will not be posted. If you find a mistake on your grade listing, please contact the course teaching assistant as soon as possible. It is your responsibility to uncover and notify the instructors of any errors.

Testing Circumstances

Because of the large-lecture nature of this course, you will take several examinations throughout the term on the dates scheduled on the syllabus. Please do not make any travel arrangements that interfere with exams as no late or make-up exams are given. If you need to miss an exam for any reason, it will be the exam that you drop as your lowest score. You cannot miss the final exam, and there are no opportunities to take it at a different time. If you have an irresolvable conflict with another course's final exam, you must see the instructor prior to spring break. During closed-book, closed-note exams, you must bring a photo ID and you are not allowed to wear headphones, to wear a hat, or to communicate with anyone in the classroom except the course instructors and exam proctors. If you have been certified as needing to take an exam under special circumstances, please see me privately.

THE ACTUAL DOCUMENT

Your course syllabus unquestionably contributes to students' overall first and lasting impressions of the course and fundamentally affects the tone of your course.[4] Especially because so many of the nonscience majors in the course come in telling themselves, "I can't do science!" it is especially important in ASTRO 101 that your syllabus convey a sense of excitement and hope for success instead of simply a list of rules and regulations. It is worthwhile asking some trusted students to review your syllabus because the overall tone that it conveys may be difficult for you to judge.

There is some basic information that any syllabus should contain:
- course name and number
- your name,[5] office location, phone number,[6] and e-mail
- scheduled office hours
- policies regarding your availability outside of office hours
- prerequisite courses or skills[7]
- required purchases such as textbooks,[8] rulers, and protractors
- policy on using or having access to calculators, personal digital assistants (PDAs), Internet, and so on
- detailed description of how grades are determined
- descriptions and goals of assignments and tests
- dates, times, and locations for all tests or other out-of-class requirements[9]
- policy on missed classes or tests

---

[4] Remember to make additional copies for students who add the course late.

[5] Although not likely to go on your syllabus, you should clearly tell students how you would like to be addressed, as many freshmen are unfamiliar with college culture and individual professors differ greatly on this issue.

[6] Some professors routinely list their home numbers and invite students to call them "any time they need help." Think carefully before extending this offer.

[7] Often, ASTRO 101 has no explicit college-level math requirement listed in the catalog, and therefore students assume that "you don't need math in this course." Explaining the level of mathematics that will be used with some examples, even when it is a very low level, can be important in establishing appropriate expectations.

[8] We often place a copy of the text on 2-hour reserve in the library for those students who wish to borrow ours.

[9] We emphatically suggest that students inform their families of these test attendance requirements to avoid the oft-heard excuse that "my parents bought me a plane ticket home that leaves before the final exam."

Additional items that we recommend be included in the syllabus or on separate documents distributed on the first day include the following:

- instructor's philosophy on roles and responsibilities
- detailed list of course goals and objectives (see Chapter 2)
- course calendar including exams, drop dates, and holidays
- an explanation of how this course fits into students' overall education and the specific university goals
- firm statement on academic honesty (often already available)
- list of appropriate additional courses for those looking to take more astronomy
- community resources such as a planetarium, museum, or amateur astronomy club
- allocated space for students to write the names and contact information for two or three classmates

Our syllabi tend to be somewhat lengthy, often six to eight pages. Because of length, some faculty opt to only have their syllabus on the Web or have the syllabus copied at the bookstore and sold to students at cost. Regardless, your syllabus should be treated as a scholarly and creative endeavor.

Chapter 5
Lecturing for Active Participation

Is it wrong to lecture? Given the substantial research showing that lecturing is highly ineffective in promoting deep learning, why should a book that purports to talk about better teaching even discuss lecturing at all? In fact, if you attend professional development workshops for faculty or read articles about effective approaches to teaching, it is easy to take away the message that lecturing always constitutes bad teaching. Lecturing, a term often used very much in the pejorative, is often characterized as nothing more than a highly inefficient method for transferring the information in the lecturer's notes into the students' notebooks, often with no mental processing by the students. The lecturer is portrayed as droning on in a dry monotone while the students, totally disengaged from the process, are wondering only, "will this be on the test?" Admittedly, if this is your image of lecturing, then there is little here to defend. And yet, faced with a class of 80 to 300 students in a large, auditorium-style lecture hall, what else can we do but lecture? The answer, we contend, is plenty. But we also contend that even in the most reformed large college classroom there will still be some—perhaps even a significant amount of—lecturing. Lecturing is not inherently a bad thing. Used appropriately, lecturing can be an important tool in instruction and is likely to remain the foundation of most large-lecture astronomy courses.

> **Active learning means engineering an environment where students' minds are engaged in critically reasoning while also monitoring their own progress toward a deeper and more connected understanding.**

To help frame this discussion, we return to the central corpus of our book: *it is not what the instructor does that matters, it's what the students do*. There is nothing inherently wrong with standing at the front of the room and speaking clearly and eloquently to your class. What is worrisome is having a room full of students absently taking notes with little or no mental processing of the information while, too often, they are thinking about things completely unrelated to astronomy. It is not active bodies we are after but active minds. As long as students are thinking critically about the concepts being presented and attempting to integrate the new concepts into their existing knowledge, then at least the potential for meaningful learning exists. There is no reason why this cannot happen during a lecture—that's the good news. The bad news is that to rely solely on lecturing is to stack the deck against achieving the kind of deep learning for which we strive. To our dismay, research repeatedly shows that teacher-centered lectures have only a limited effect on student learning. For example, one classic study reported that only 66% of students showed signs of attention to lectures after 18 minutes and no students were completely attentive after 35 minutes (Verner & Dickinson, 1967). All too often in our own classes, we look out into the crowd and find students reading newspapers, talking with one another, and, at times, blatantly sleeping with their heads on the tables. But more disturbing than a lack of paying attention during lectures, research suggests that student learning is poor even for those students who pay attention in class. In the context of reviewing learning gains of 6000 physics students, Hake (1998) reported that students attending lecture-based courses had far lower cognitive gains than those who attended classes characterized by some form of active-engagement.

With all this negativity, why lecture at all? In fact, there are a number of things that can you can do very effectively by lecturing:

- *You can communicate your enthusiasm for the subject.* As scientists, we do what we do because we have an insatiable curiosity and love for our subject. Lectures provide an opportunity to share the human side of science with students who, raised on media images of scientists, too often see science as a sterile and inhuman endeavor devoid of the kind of personal creativity and passion that students normally equate only with the arts and humanities. If nothing else, displaying openly your personal passion for astronomy provides an opportunity to keep students motivated to read and complete assignments.
- *You can present the newest, cutting-edge discoveries.* One of things that generally sets astronomy apart from most other sciences is the ravenous public enthusiasm for our latest discoveries (paleontology would be one notable exception). Though your students might not yet be at a point where they can fully appreciate the implications of the work, sharing recent discoveries creates a sense of excitement and interest. And, because you explain the discoveries in simple terms,

students have something to share with friends and family. Whereas students are unlikely to talk to friends about how the H-R diagram can be used to estimate stellar distances, they are apt to talk about what they learned about the latest space mission. For nonscience majors, this opportunity to share with others can be an empowering experience.

- *You can share the stunning images that astronomy affords.* Astronomy is, or at least can be, a visually engaging subject and, every time you get to show your class the Hubble Deep Field image, aren't you thankful that you don't teach calculus to

> *Excellent Resource*
> A great source of images and information is the Astronomy Picture of the Day at http://antwrp.gsfc.nasa.gov/apod/astropix.html. This site includes a searchable archive.

nonmajors? Many of the images that we show are in our students' textbooks, but these small textbook images do not have the same impact as when presented on a large screen accompanied by an enthusiastic description in the context of what is being studied.

- *You can emphasize the important points from the textbook to help frame and guide students' reading.* Typical astronomy textbooks are full of information, but students have immense difficulty separating the details—those things that are interesting to read but not of critical importance—from the central message. For instance, isn't the primary message about the interstellar medium (ISM) its very existence rather than the details of its composition and the various means we have for detecting it? Lectures are ideal for providing needed focus and communicating what aspects are truly important.

Our message is not that you should avoid lecturing at all costs; rather, when you do lecture, you and your students should know why you are doing it. Most important, you should try to avoid uninterrupted lecturing (i.e., talking directly to the class) for more than 15 to 20 minutes without taking a short break to engage in a student-centered activity such as individual questioning, think-pair-share, talk to your neighbor, or in-class writing, which are briefly described in the following sections.

INDIVIDUAL QUESTIONING

We hardly claim to be offering new advice in suggesting that you should be asking your students questions during class. However, we often find that faculty in large auditoriums pose rhetorical questions that students have learned they needn't answer—if they wait just a moment, the professor will answer it for them. The effectiveness of asking questions to produce thoughtful answers can be greatly improved by following a few simple guidelines that many novice faculty forget:

- *Avoid using only low-level questions*—Engaging questions require students to reason and apply, not just to remember. As a means of increasing class interaction, there is nothing inherently wrong with asking, "Does anyone know what this is called?" but questions like this do not promote meaningful learning.

Box 5.1 Questioning Tips

- Avoid only using low-level questions
- Avoid questions that telegraph the answer
- Allow sufficient "wait time"
- Avoid responding positively to the first answer
- Don't answer your own questions
- Plan your questions ahead of time
- Retract poor questions
- Solicit answers from around the room
- Avoid nonspecific feedback questions

- *Avoid questions that telegraph the answer*—Some questions, by the way they are asked, can be answered without having to understand the underlying concepts. An example from everyday life is, "Are you actually going to wear that shirt with those pants?" Clearly, you don't need to be a fashion expert to answer this.

- *Allow sufficient "wait time"*—This is perhaps the single most important advice about classroom questioning techniques, and is undoubtedly the most difficult to follow. Students need time to process a question, reason about the answer, reflect on their reasoning, and formulate a response using the vocabulary of the discipline. In a widely cited study of over 300 classroom tape recordings, Rowe (1974) reported a mean of only one second for teachers remaining silent after asking a question. Further, she found that an average of only 0.9 seconds elapsed between initial student answers and the teachers' responses. Rowe and others (viz. Tobin & Capie, 1980) found that increasing these wait times to between 3 and 5 seconds provided significant benefits including more thoughtful student responses and broader participation by all learners. We suggest that, when you pose a question, tell the students how long they will have: "Take 20 seconds and think about this by yourself...." Then, and this is the really difficult part, look at a watch or count to ensure that you really do allow them the time you promised.

- *Avoid responding positively to the first answer*—When one of the students immediately answers correctly, it is temping to say, "That's right. Did everybody understand that?" The effect of this is to shut down thinking in the rest of the students. Often, it is better to respond,

"Are there other ideas?" If your class gets used to your initial responses being similar for both correct and incorrect responses, they will continue to think and reason beyond the first response from another student.

- *Don't answer your own questions*—If you get in habit of answering your own questions, students stop thinking about the questions and just wait for you. Especially at the beginning of a semester, students are particularly reluctant to answer, and it can take real patience to outwait them. However, it has been our experience that, given enough time and a long enough pregnant pause, they will answer or pose a question themselves.

- *Plan your questions ahead of time*—There is certainly nothing wrong with asking students that insightful question that just comes to you during lecture (and, if it works, writing it down for future use). However, developing good questions, ones that challenge without overwhelming, takes practice and is best not left to chance. Perhaps the best source of questions is old exams for which you have the data (item difficulty and discrimination if it was a multiple-choice question) to judge whether the question is at an appropriate level.

- *Retract poor questions*—If you ask a bad question—we do this all the time—rather than answering it, just say, "I'm sorry, that was a horrible question. Let me try to phrase it another way...."

- *Solicit answers from around the room*—It is incredibly gratifying to have that student, normally in the front row, who always comes to class, always knows the answers, and loves to share. However, just like when you answer your own questions, students can learn to wait for Chris to answer rather than thinking for themselves. Make sure to call on students from around the room and, if needed, chat privately with Chris after class about the need for giving others the opportunity to answer.

- *Avoid nonspecific feedback questions*—Questions like, "Does everybody understand?" and "Is that clear?" often lead to a lot of nodding heads but seldom do they provide a true measure of how well the class is really understanding. The most desirable approach is to ask a question that directly assesses student understanding of the concept being discussed. To determine if students understand the concept, questions like, "What are some examples of this idea?" or "How would increasing the temperature change this process?" provide you with a much clearer picture of student comprehension. A more general approach that at least improves upon "Does everyone understand?" is to ask students to talk briefly with their neighbors about any points of confusion and then ask them to share any questions that they cannot resolve.

THINK-PAIR-SHARE

This simple questioning technique, introduced by Lyman (1981) and popularized within the physics community by Eric Mazur in his book *Peer Instruction* (1997), provides an additional layer of structure beyond simple questioning and takes advantage of the power of discussion for making meaning. You begin by asking students to think quietly about a question related to material that has just been discussed (the amount of time can vary a great deal depending on the nature of the question) and commit to an answer, preferably in writing[1]. It is critical that they commit to their initial ideas or else they will say, "Yeah, I already knew that," and not actually resolve conflicting information.

You then ask students to discuss their answer and, more important, their reasoning with their neighbors or in small groups. It is during this period of sharing and open critique that students can come to more sophisticated understandings. This results both from the focus that discussion often brings to an issue and from the ability of students with a better understanding to explain—in the process of defending their own answers—in a language that is more meaningful to other students than anything we, as professors, can hope to achieve. Finally, have students share their reconsidered answers by shouting out (the simplest method), voting, or simply volunteering to summarize different explanations in a class discussion.

Think-pair-share allows a great deal of latitude in terms of the formality with which you use it. At one end, you could ask an open response question like, "Will our Sun ever produce a nova? Why or why not?" You first allow your students to think about it individually for a moment and then you say, "Okay, talk to your neighbor about your answer." Finally, you ask, "Who would like to share what the person next to them thinks?" and, after gathering several responses, wrap up with a brief discussion. At the other extreme, you could pose a multiple-choice question and have the students enter their initial responses using a wireless student response system in which they have identified themselves so that they receive participation credit. Following their group discussions, students key in their revised answers and then you display the results on a

---

[1] Numerous questions are available in *Peer Instruction for Introductory Astronomy*, by Paul Green, Prentice Hall, 2002.

computerized projection screen and discuss the correct answer. After class, you review the students' response rates before and after discussion to evaluate the usefulness of a particular question. (You check whether a sufficient number of students had difficulty before discussion to validate the need for the question—about 50% is an oft-used value—and whether enough students changed to the correct answer as a result of discussion to confirm the discussion's usefulness.)

There is of course a great deal of room between these two extremes, and most us do not (dare we say "yet?") teach in classrooms equipped with such sophisticated wireless response systems. Although the low-structure approach can be valuable, we suggest that you consider the latter approach as the ideal and then think about ways that you can incorporate elements of this approach with the facilities you have to work with. When data are collected and displayed, individual students assess their understanding relative to the class. Further, this provides the instructor with a snapshot of where the class is as a whole. There are ways to get this kind of feedback without high-tech support. The simplest, and one that we often use, is to have students vote by raising their hands. As you might anticipate, making this work effectively requires developing a classroom culture in which students are not overly worried about being wrong. To make the vote less public, students can vote with their eyes closed or hold up small cards (3×5 index cards work well) with A, B, C, D, or E printed on them. In these cases, it is important for the instructor to scan the room and report roughly how many students are giving each answer.

TALK TO YOUR NEIGHBOR

In some ways, this is just a less formal and more easily implemented variation on think-pair-share in which the step of individuals initially committing to an answer is omitted. What makes it different is the types of questions you pose. If you are asking content questions, we suggest that the step of considering the question individually is crucial and should not be omitted; talk to your neighbor should be used for more reflective questions. Consider the difference between just asking the class, "Does everyone understand?" and instructing them, "Turn to your neighbor and discuss what you think is the most difficult concept from today's class." The first question generally leads to a lot of absent nodding by students, which research repeatedly shows is no indication of student comprehension. When given enough time, the second approach often leads to a rich discussion that can prompt insightful questions from the class because, as a result of the discussion, points of confusion are shared with peers, which makes talking about misunderstandings less threatening.

Box 5.2 Examples of Talk-to-Your-Neighbor Questions

- Where else in the course have we seen this concept applied?
- What are three real-world applications of this concept?
- If you were trying to explain this idea to a friend, what part would be most difficult?
- What are all of the different variables that impact this situation?
- What is the most interesting question left unanswered by today's class?

IN-CLASS WRITING[2]

We would be the first to advocate that, if resources permit, an intensive writing component ought to be included in all college courses. Not only does frequent writing help reinforce one of the most important skills that students will take into the workforce, but it also fosters deep learning by forcing students not just to consider challenging concepts but to explore ways to structure the important components of concepts in their own words. Ideas held fuzzily in the head become more organized, refined, and sharpened when expressed in writing. However, for most large-lecture ASTRO 101 courses, the expectation that instructors would collect, grade, and provide meaningful feedback for a series of multipage papers is unrealistic. However, the simple technique of the in-class writing (ICW) activity provides students the opportunity to write reflectively and, at least at an elementary level, receive some feedback on their writing and ideas. There are many ways to adapt this particular strategy to fit local conditions. We describe our implementation as one of many.

Box 5.3 Two ICW Question Examples

General: Describe the life story of our Sun from age 0 to 20 billion years; provide as much detail as you can.
Specific: You observe Cepheid variable stars in two difference galaxies, A and B. The star in galaxy A has a period of 10 days while the one in galaxy B has a period of 14 days. The star in galaxy A appears brighter than the star in galaxy B. What can you conclude, if anything, about the relative distances to the two galaxies? Explain your reasoning.

Our in-class writing activities are unannounced. At some point during the lecture a slide appears with a title like, "ICW #7," which the students know means they should take out a piece of paper and write their name on the top right

---

[2] We would like to acknowledge Phil Gaines, Director of Composition, Montana State University, as the person who first introduced us to this technique.

corner. Below the title is a question, usually related to the material we were just discussing, on which they are expected to write about half a page in response. Questions range from quite general to very specific (see examples in Box 5.3) and usually require no more than five minutes to complete. Part of building a risk-free classroom culture around this kind of writing is continually reminding students that the purpose of the exercise is to use writing as a tool to help solidify their understanding; it is not to test their writing skills. We wrap up the activity by having several students read their narratives aloud to the class. It sometimes takes patience to get a volunteer, but as long as students don't feel the risk of being ridiculed, someone usually offers to share. If not, we just crumple up a piece of paper, face away from the class, and toss it over our shoulder to select the lucky volunteer—simple but effective.

It is important to emphasize that students read exactly what they have written, not paraphrase, and ensure that the rest of the class can hear. We routinely offer the direction, "READ LOUDLY AND WITH FEELING," presented with tongue firmly planted in cheek, which gets the job done. We always thank the reader for sharing but avoid making specific comments about the content.

After we have heard from a few students, we ask the rest of the class for comments:

> **You don't need to implement all these things at once. We recommend starting with one or two and slowly expanding your repertoire.**

- "Is there anything else you would add?"
- "What did the passage make you think of?"
- "Were there any specific points that you heard with which you disagree?"
- "Has anyone written something from a very different angle that she or he would like to share?"

After allowing sufficient wait time, we usually offer a few comments of our own and then have students pass their papers along to be collected. Because there will always be some students wanting to make corrections as a result of the discussion, we reinforce that any honest effort will receive credit (we use pass/fail) and that we will only read a few papers carefully. We do go through the papers to record who participated meaningfully but do not "grade" them. We do read some more carefully and offer constructive comments. It has been our experience that the small fraction of papers that don't represent an honest effort are easily spotted. These students are not given credit but are invited to correct their work, or

write a new answer at the bottom of the
same sheet, and resubmit within one
week for full credit.

Excellent Resource
Great Ideas for Teaching Astronomy,
Stephen M. Pompea, Brooks/Cole
Publishing, 2000;
ISBN: 0534373011

The preceding list of short
active learning exercises is certainly
not exhaustive and, once in the habit of
trying to do something different every 15 minutes to break up the lecture,
other ideas will no doubt emerge. For now, the important message is never
to plan a 50-minute lecture. Instead, plan a series of 5- to 15-minute mini-
lectures (20 minutes, if you must!), each with its own learning objective and
with a clear beginning, middle, and end. Active learning exercises are then
used in between. Of course, during a complete lesson, you might present
several of these, and they should be linked together logically, but, in a
sense, they should also stand alone. In fact, aside from increasing the
likelihood of student engagement, this strategy has the advantage of
increasing your flexibility in that, if you get behind because of the need to
address a particular student difficulty or expand on an issue of interest, it is
easier to eliminate just one mini-lecture than to speed ineffectively through
a planned 50-minute lecture.

PLANNING LECTURES

As you sit down and think about what you want to do for Monday's class,
the most important question to ask yourself is, How do I want my students
to be different as a result of this hour we spend together? This is essentially
the same as asking what you want them to learn except that it focuses your
attention on higher-order learning objectives; you can't answer this question
with, "I want them to know the definition of ...." It is important to identify
a small number of (possibly only one) learning objectives and the
approximate cognitive level of those objectives for each mini-lecture. You
don't need to be explicit about the cognitive level but it is important to
know whether you are aiming for the lower or higher levels because this
affects the amount of time you must spend and, for higher-level objectives,
might even suggest that a strategy in addition to lecture is required.

Once the learning objective is selected, decide what specific things
you would like students to have in their notes at the end of the lecture and
be sure these words or figures are written clearly and that students will
understand the importance of writing them down. Next, outline the general
"story" you want to tell. This might be a true historical tale such as Adams
and Leverrier's attempt to predict Neptune's location, or it might be a
concept that you wish to develop, such as the relationships among

brightness, luminosity, and distance. In either case, the "story" will have an introduction, which often involves a quick review of key concepts; a middle in which the key ideas are explained; and a conclusion, in which the significance or application is emphasized.

> *Excellent Resources:*
> A *History of Astronomy from 1890 to the Present,* David Leverington, Springer, 1995; ISBN: 3540199152
> *Cambridge Illustrated History of Astronomy,* Michael Hoskin (Ed.), Cambridge University Press, 1997; ISBN: 0521411580

It is a good idea at this point for you to review the information in the students' textbook so that you know what they can find directly in their text and what you are going to provide as a supplement. Sometimes, you will disagree with what is in the textbook, but unless you are simply updating the information with very recent discoveries, we recommend keeping your criticism to yourself. In general, students react negatively to loud proclamations of errors in their books, which cause them to mistrust the remainder of the book and wonder why you had them buy it in the first place.

Most available astronomy texts are quite comprehensive, and you will generally find that the concepts you are addressing are covered. What you provide through lecture should be a framework for students to study the text. Students appreciate direct references to what they can find in their text such as, "You don't need to copy this down; just note that this is on page 338 of your text." Perhaps more important, they need guidance as to what detail they can ignore—"You will find a lot of detail in your text about the alternate shell and core burning that takes place during this phase, but the important point to remember is that the star grows larger and the surface cools." You might also consider insisting that students bring their books to class and encouraging that they write in the margins.

In addition to knowing what information the students have available in their text, it can be very helpful to review a couple of other texts for their treatment and at least one comprehensive historical reference to learn those brief stories or other facts with which to embellish your lecture to make it

> **Many textbook companies provide CD-ROMs to faculty with all digital images from the text and video clips—but only upon request!**

more interesting. How much of this material to use is both a matter of judgment and a balancing act. If you spend too much time on extraneous information that does not directly address the learning objectives (i.e., is not on the test) students tend to tune out; they judge what's important by what you test. On the other hand, student attendance will drop dramatically if all you do is "read the book" to them and do not take the opportunity to make the subject more human.

Finally, gather all of the images or other media, such as sound recordings (pulsars, for instance), that can help supplement and enliven your lecture. We are advocates of electronic presentations (Microsoft PowerPoint in our case) because of the flexibility they provide. However, most of what we are describing can be accomplished with a mix of slides, overheads, and notes on the black (or white) board. Arrange your images to tell the story and present in words only the most critical points that you want students to copy in their notes exactly as you have them. Students often believe that their role in class is that of scribe; it is important to emphasize to students that their role should be learning, not copying. Even if you tell students that they needn't write down everything you put on the board, there will always be those that think they will be missing something if they don't and therefore you should be intentionally succinct. If

> **Be careful how much you write on your overheads or slides...you can type slides so much faster than students can copy them down that it is easy to project far too much information at once.**

you are going to present anything too complicated to copy down, we suggest instructing students that it is not critical (but then, why are you showing it?) or, better, telling them where they can access it (either in their textbook, on reserve at the library, or on the Internet). Some faculty find it useful to provide copies of their lecture notes to students as photocopies or on the Internet. Personally, we haven't come to any resolution on this idea as one of us does and one of us does not. The overwhelming advantage of providing notes is that students can bring them to class and make notes on your notes. Unfortunately, the overwhelming disadvantage is that many students believe (and are all too often correct in many classes) that if they have the lecture notes, they do not need to attend class at all! Unless you have significant amounts of time, it is probably more useful to spend time designing interactive lectures than creating comprehensive Web sites.

With all of the major decisions made, what remains is to plan the details. The amount that you write down in your personal lecture notes about this is really a matter of personal preference. We suggest that the first time you teach a class, it is better to err on the side of overly detailed lecture notes (to refer to, not to read!) than to go in unprepared. However, with experience, we find that our PowerPoint slides and images themselves provide the necessary structure.

What we ultimately find more useful than spending time on lecture notes are our own reflections on what worked and what didn't about a particular lecture. It does require some discipline, but if you can get yourself in the habit of sitting down and writing in a course diary about the strengths and weaknesses of each class (while those ideas are still fresh), you will find it to be an invaluable resource both to ensure that you repeat successful elements (even the jokes that worked) and make changes to avoid what didn't work.

Chapter 6
Implementing Small-Group
Collaborative Learning

Interactive lecturers who successfully engage students in questioning, ask students to explain their reasoning, and require students to compare their ideas with other students obtain enormous pedagogical advantage over faculty who merely talk at their students the entire class period. However, interactive lecture strategies that preserve the professor at the center of the learning process still fall short of our basic premise—*it is not what the instructor does that matters; it's what the students do.*

Probably the most widely researched cluster of teaching strategies that place students at the center of the learning process involve small groups of students working collaboratively to solve complex problems requiring the intellectual resources of the entire group. In the literature, these strategies go by names such as "collaborative learning," "cooperative small-group learning," and several other variations; although variations in titles are meant to communicate subtle differences in technique, these strategies have a great deal in common and are collectively the genesis of the group approach we employ. In our classes with nearly 200 students, we challenge groups of three to four students working together to solve a problem, complete a task, or create a product.[1] The approach is based on the idea that learning is a social activity in which the students talk among themselves and it is through their social interactions that learning occurs. By presenting our "in-the-trenches" experience with collaborative learning groups, it is our hope that readers will

> **Excellent Resource**
> The National Institute for Science Education has established a comprehensive Web site dedicated to collaborative group learning at the introductory college level (URL: http://www.wcer.wisc.edu/nise/CL1/CL/clhome.asp).

---

[1] Much of the material appearing in this chapter has been adapted from "Learning Through Sharing," *Journal of College Science Teaching*, 31(6), 384—387 (2002), by Jeff Adams and Tim Slater.

be able to use what we have learned as a starting point for implementation of some form of participatory group work in their own courses.

NUTS AND BOLTS

In the fall semester of 1997, we introduced learning group activities into both sections of our introductory, one-semester astronomy course, which students generally take to fulfill the university's general education requirement. With no materials yet developed, we took the bold step of announcing on our course syllabus that students would be doing at least one group activity each week and that these would collectively contribute 25% toward their course grade. During that semester, we developed and field-tested a series of highly structured, 30-minute duration, collaborative group activities, which students work on in groups of four (see an example of one of these structured activities in Appendix A). Each group submits just one completed activity sheet—something that greatly reduces our grading burden—and students share the same grade for their work.

Box 6.1: Examples of Unstructured, Open-Ended, Collaborative Group Learning Activities

Our structured activities [see Adams and Slater (1998, 2000) and Appendix A] usually involve two to six pages of worksheets. In contrast, these unstructured activities are presented in a paragraph.

Galileo's Observations. Your group should select what it believes to be Galileo's single most important astronomical observation, why it was most important, and explain what he observed using sketches.

Tourist Attraction or Sacred Ground. Your group has been asked to arbitrate a dispute between a tour bus company and a nearby Native American tribe. The dispute surrounds an ancient medicine wheel recently discovered by a team of university archeologists. Using sketches as necessary, compose a legal brief that describes what a medicine wheel is designed to do astronomically and summarize the opposing positions of the two groups.

Cost of Bathing on the Moon. The anticipated cost of transporting a gallon of water from Earth to the Moon is $15,000. Estimate the cost of taking a single-day's supply of water for your group to the Moon by determining how much water each of the group members use in a single day.

Evolutionary Sequences. As a plot of luminosity versus temperature, the HR diagram is useful for describing how stars evolve over time even though "time" is not the label on either axis. As a group, create an imaginary graph of "dollars of financial income" (vertical axis) versus "weight" (horizontal axis) and use it to describe the past and future life cycle of one of your group members. Clearly label your diagram and provide a figure caption clearly explaining each life phase.

(A longer selection of classroom-ready examples is provided in Appendix E.)

The activities were designed to maximize interactions among students by focusing on open-ended questions that engender student discussion. For example, our first activity challenges students to *use a telescope catalog to decide how best to spend $6000 to equip a community astronomy program.* Using the vocabulary of education, such an activity has multiple entry and exit points and the task can be approached at varying levels of complexity based on the student population. (Using this activity at a recent faculty teaching workshop, we found one group of participants spending their money on PVC pipe and mirror blanks—not the typical entry point for our introductory students—yet still intellectually engaging.) As another example, we provide students with 18 photographs of galaxies and ask them to devise and defend a classification system for the collection. Certainly some students naturally adopt a similar tuning-fork system to what Edwin Hubble advocated, but many other schemes are creatively presented. This is one of several process-oriented activities designed to involve students in a more authentic scientific task than would normally occur in the lecture environment (viz. Adams & Slater, 2000).

On each activity task sheet, we list four blanks for student names next to the four roles we ask students to fulfill; the roles rotate with each new activity. The task of the *leader* is to be sure that each member of the group contributes, that everyone's ideas are represented, and that the group stays on task to finish the activity in the allotted time. The role of the *recorder* is to write the group consensus answers on the answer sheet and to be sure that the assignment is turned in before the class is finished. The *skeptic's* role is to ask the questions, "Are we sure?" and "Why do you think that?" The *explorer's* task is to try to investigate ideas and areas that no one else has considered. When there are only three members of a group present, we suggest that students forgo the *explorer's* role. Students print their names and sign the sheet with the understanding that signatures not only certify their own attendance but also that of the rest of the group. Spot checks have never uncovered a group filling in the name of a member not in attendance.

We aggressively enforce a rule of four students to a group. This is because if one member is absent, there is still a group of three. If a group has only two members, then the likelihood that they will need help from the instructor seems to go up—something well worth avoiding in a classroom of 200 students. An unavoidable problem is that in any large class some students are bound to arrive late—sometimes for very good reasons (and sometimes not!). One approach to dealing with this is to require students arriving late to meet at the front of the class and form new groups. This avoids the problem of latecomers getting credit for the work already completed by the group.

Another particularly interesting collaborative learning strategy well suited to classes of around 30 students is *jigsawing*. The approach is to divide students into four or five expert groups. These students become experts on a particular area of responsibility. The class is then reorganized into four or five project teams that are composed of at least one member from each expert group. It is the responsibility of the experts to teach their assigned project team members whatever they need to know to solve a problem. For example, imagine tasking project teams to create a written proposal for constructing a 10-m telescope at one of a number of pre-selected mountaintop observing sites. You might first create separate expert groups who learn how to evaluate weather patterns and seeing, calculate actual travel expense projections for visiting astronomers, determine local cultural and environmental issues, evaluate the local economy and availability of technical workers, and conduct seismological risk evaluation. These experts would then need to explain how each of these issues impacts an observing site selection to their project team.

LESSONS LEARNED

Successfully implementing student learning groups requires a certain level of "buy in" from the students. We explain to the students, both in the lecture and in the syllabus, that we are using collaborative learning groups to allow them to be more actively involved in their own learning and that we believe that they will learn more from one another than they will from our lectures alone—no matter how entertaining we personally think we are. We also tell them that although we love to lecture, we believe that allocating precious class time to learning group activities is worthwhile. We repeat this message frequently throughout the semester.

The most inexact part of collaborative group approaches involves how the groups are composed. Teacher-assigned grouping is a formidable task in the large-lecture course from the management perspective alone, not to mention the possible social *faux pas* waiting to erupt in any college environment. In our classes we have the students self-form groups with the advice that groups will function most effectively if there is a common level of interest and class attendance style. Based on the results of focus group interviews, we suggest that nontraditional age students should be encouraged to work together. At the beginning of the first few activities, we encourage the groups to engage in honest discussions about levels of commitment to learning astronomy and suggest that those not comfortable with their initial group feel free to switch. The rare group meltdowns that occur later in the course usually seem to occur for one of two reasons. Either one group member wants to go slow and understand every part of the learning activity while the others want to get to a final answer and finish as soon as possible, or fellow group members repeatedly fail to attend class on activity days. In these extreme cases we allow students to change groups. We also seem to encounter about one student each semester who feels very strongly that she or he should not be "held back" by other students in the group who are unable to keep up or understand. If a frank discussion of the value of teaching their classmates as an aid to deeper understanding for themselves makes no impression, we regretfully permit the student to work independently.

Maintaining a sense of group identity is not always easy in a class where significant numbers of students are still adding and dropping the course three weeks into the semester. In the past, many students have appeared to work with whatever group was nearby on that particular day. More recently we have been using assigned group numbers to help promote a sense of stability. During the third activity of the course, each group is given a sheet of paper with a different group number to record the names of the group. They then record this same group number at the top of all future activities. We feel that this has engendered a sense of stability in the groups and has provided an additional bookkeeping benefit. By entering the group numbers into the spreadsheet in which grades are kept, the class list can be sorted by group number, making the recording of activity scores much faster than before.

In focus group interviews conducted during the first year, we found that many groups were not rigidly adhering to the assigned roles, with the exception of the *recorder*. Accordingly, we now remind students regularly that roles are used because, due to the limited class time, there is not an opportunity for the natural group roles of their members to emerge, and rotating roles allows everyone to participate.

The first barrier that many faculty see to implementing collaborative group work in the large-lecture environment is the seating arrangement. In a workshop, Michael Zeilik encouraged us to just go ahead and "do it" with the assurance that the students would find a way to make it work—in general, he was right. Although many students sit in their seats, some groups sit on the floor in the aisles or at the front of the room or sit outside the room in the hallway. However, we did discover that for those staying in the seats there was something we could do to help. Our lecture hall is composed of long tables running from one side of the room to the other. When students attempted to work by sitting four across, invariably one or two people were left out. We now strongly urge students to sit two in one row and two in the next row in the shape of a square so that they can work together more easily.

One early idea we used to encourage students to work together was to provide only one copy of the task instruction for each group. Due to significant photocopy costs, we now ask each student to purchase a set of activities (CAPER Team, 2002), so we do not know if such a one-copy approach makes any substantial difference in encouraging more meaningful group interactions.

FINAL THOUGHTS

In this approach we have not abandoned lecture entirely. We estimate that 70% of our class time is still allocated to interactive lecturing. Furthermore, we do not find that this technique in itself is a magically successful approach to teaching large lectures for nonscience majors. However, by focusing on engendering an environment where students are active learners, we have made significant improvements in our course reputation, as reported to us by students and those in the General Studies advising office. Although attendance is never 100%, as we would wish, our attendance is usually in the 80% range, which is quite high for a general education course at our institution (if you are surprised by this, we strongly encourage you to do a head count at your next class meeting). To encourage students to attend class even more often, we do not announce beforehand on which days we will have activities—something students do not complain about on course evaluations.

Probably most important, we now interact individually with our students much more often than we did before as we spend our time going from group to group asking probing questions. In the end, we spend more time working directly with students and, in turn, students spend more time actively engaged in their learning.

Chapter 7
Strategies for Writing Effective Multiple-Choice Test Items

In teaching hundreds of students in the large-lecture sections of ASTRO 101, we have been forced to rely heavily on machine-scorable, multiple-choice exams. We fully acknowledge that it is difficult to probe our students' conceptual understanding using multiple-choice exams. Certainly, essay questions, or possibly one of several alternative assessment strategies available (see Chapter 8), would be preferable for assessing the strengths and weaknesses in both our students' learning and our instructional approaches. However, given the minimal grading support we have available and other demands on our own time, the use of multiple-choice tests is mostly unavoidable. Does this mean though that we have reduced our assessment to the lowest common denominator—the bubble sheet? The answer, fortunately, is not necessarily. Though it is easy to create multiple-choice questions focusing on lower-level knowledge and comprehension— and there is nothing wrong with having some of these questions in any test—it is also possible, with some effort, to ask multiple-choice questions that probe student understanding at more advanced cognitive levels. Ultimately, we do not advocate relying entirely on multiple-choice tests as the only means of assessment but, given some simple guidelines, it is possible to create multiple-choice tests that provide the instructor and students with meaningful data that go well beyond testing simple vocabulary.

GENERAL CONSIDERATIONS

Probably the most difficult perspective to adopt about the difficulties in creating effective multiple-choice tests is that, despite outward appearance, these tests are actually inherently nonobjective. In particular, we often overhear colleagues talk about how scoring essay exams is very subjective as compared with multiple-choice exams (with scoring of numerical solutions to solved problems falling somewhere in the middle). In fact, our colleagues are only partially correct—the "scoring" of multiple choice exams is completely objective in that all faculty would grade responses identically as right or wrong. However, question structure and wording are undeniably subjective, and these choices can greatly influence a student's success on a question independent of his or her level of understanding.

Let us illustrate by means of an example. If you were to ask 10 faculty each to write a multiple-choice question to assess student understanding of Kepler's third law, you'd likely get a very wide variety of proposed questions, which would lead to greatly varying degrees of student success in answering the questions. We argue that this is inherently nonobjective or, dare we say, subjective. Compare this with asking the same 10 faculty to grade student essays responding to the prompt, "Explain Kepler's third law." We propose that although there would be some spread in the grading results, it would be much less than the spread in grades from the 10 different multiple-choice questions. We are not arguing that multiple-choice tests are inherently flawed but rather that their objectivity should not be overestimated.

The ultimate goal of testing is to measure what the students actually understand, and the process of interpreting the meaning of a student's response to a multiple-choice test is a subjective one. That being said, the following guidelines are designed to maximize the likelihood that your questions will indeed measure student understanding of critical content and not be contaminated by other effects.

There are three major issues surrounding creating, and revising, the most effective multiple-choice questions possible: (1) the physical format and layout of the questions; (2) the conceptual hierarchy of the questions; and (3) the statistical item analysis. We offer some general guidelines related to each of these issues that we have found useful in guiding our own test writing. Like all generalizations, they are intended to serve as a guide, not a set of unbreakable rules. It has been our experience that, unless we can justify clearly why we are violating one of these guidelines, it is wise to consider revising the item to conform.

LAYOUT

The first issue deals with the overall structure of the question, which is often called an *item*. In discussing structure, we distinguish between an item's *stem*, the opening text that sets up the question, and an item's *responses*, the collection of statements or values from which the student must chose the correct one. The responses comprise the correct answer and a collection of *distracters*, which are reasonable-sounding incorrect responses. There is no rule about the number of responses that must be supplied. Most electronic forms have room for up to five, although 10-response forms are available. However, just because there are five "bubbles" available does not mean that they must, or even should, all be used. Good distracters are ones that sound reasonable, are clearly incorrect to someone who understands the concept being tested, and therefore might reasonably be selected by a student who does not understand. Consider the following example:

> You forget that the star Betelgeuse is a red giant and apply the method of spectroscopic parallax to determine its distance. The true distance to Betelgeuse is actually
> A. closer than you calculated.   B. farther than you calculated.

It might be tempting to add distracters, but there is nothing gained by adding response "C. none of the above," (or worse, "D. green.") Arguably, the response "the same as you calculated" might be appropriate; this is a judgment call. In general, though, if you can't think of good additional distracters, it is better to stop than to add ones that students will clearly see as wrong.

An important overall consideration is to create items that are testing students' understanding, not their reading ability. Long passages of text cause slow readers to skim and often miss critical details. Of course this must be balanced against the desire to ask rich questions in new and interesting contexts, including the necessary clarification to make the question clear. Certainly, any question that is particularly long compared with the average should be reexamined to see if there are places where it could be shortened.

It is also important to recognize that it is nearly impossible to anticipate all of the ways in which a question might be misinterpreted and adding all of the necessary qualifiers can make the question unreadable. This is why it is so important to build a low-risk culture in the class that allows students to feel comfortable asking questions of clarification during a test. In this way, you avoid adding those extra words to clarify a question for the one student who might worry about that issue. For instance, to eliminate all of the possible questions from the previous example, it might be rewritten as

> You forget that the star Betelgeuse is a red giant (a very luminous star in top right of the H-R diagram with relatively low surface temperature) and

apply the method of spectroscopic parallax—a comparison of a star's apparent magnitude, estimated from the H-R diagram, and its absolute magnitude—to determine its distance from Earth, which can be considered the same as its distance to the Sun because the Earth-Sun distance is negligible given the scales involved. The true distance from Earth to Betelgeuse is actually…

In an admittedly over-the-top attempt to be completely clear, the stem has actually become more difficult for most students to understand.

Box 7.1 Other Guidelines for Multiple-Choice Items

Construct items so that students can know basically what a question is probing by reading only the stem. For instance, try to avoid using the stem construction "Which of the following statements is correct?".

Avoid repeating words or word forms from the stem in the correct response such as
If you analyze the light <u>emitted</u> *from a low-density object (such as a cloud of interstellar gas), which type of spectrum do you see?*
A.   *dark line absorption* ~~spectrum~~
B.   *bright line* <u>emission</u> ~~spectrum~~
C.   *continuous* ~~spectrum~~

Related to the need for readable structure, unnecessary vocabulary can unintentionally obscure the concept you are testing for. As an example, consider two questions about what energy process fuels the stars: *The thermonuclear reactions in a stellar core are (A) fission, or (B) fusion* versus the very different item that asks *The energy process that powers the Sun is (A) heavy elements dividing into lighter elements, or (B) lighter elements combining into heavy elements.* The former focuses on testing student vocabulary whereas the later focuses on student concepts. As the instructor, your need to have a clear idea about which is most important and, if both are, you need to include items of both types.

Over the years, students have learned that when novice faculty include choices such as *all of the above* or *none of the above*, these are frequently the correct answer. Similarly, students have learned through experience that the longest of the choices is often correct and that answers *A* and *B* are more common than *C* and *D*. You can avoid these common pitfalls by making sure that all of your choices have similar lengths, similar amounts of scientific vocabulary, and that you ensure equal numbers of *A*, *B*, *C*, and *D* being correct.

Box 7.2 Contrasting Multiple-Choice Items

Are you testing conceptual understanding or reading ability? *Shorter stems and even shorter response choices are preferable. Emphasize important words such as not, except, and always in all caps.*
Less Desirable: The Stefan-Boltzmann law describes the relationship among the variables of a star's luminosity, surface area, and temperature. Consider two stars, Alpha Orion and Beta Orion, which have different colors and, of course, different temperatures. If they were to have identical luminosities, which they do not, which of the following would NOT be true?
(A) The two stars, although being at different distances and having different apparent magnitudes, would have the same absolute magnitude.
(B) The two stars, having the same luminosity, would have different peak wavelengths and, correspondingly different temperatures.
More Desirable: If two stars have identical energy output, yet one star has a higher surface temperature, this star must ALSO be
(A) smaller.   (B) brighter.   (C) cooler.   (D) closer.

Do you give clues to the correct answer in your question? *Items need to be reviewed carefully so that they probe student understanding rather than students' ability to eliminate distracters.*
Less Desirable: Which galaxy do we live in?
(A) Andromeda Nebula   (B) Local Group
(C) Milky Way Galaxy   (D) none of these
More Desirable: The Large Magellanic Cloud is a
(A) distant galaxy.   (B) satellite galaxy.
(C) spiral galaxy.   (D) Milky Way galaxy.

CONCEPTUAL LEVEL

The preceding example on fusion highlights the second consideration for test item development—be certain your items do not dwell on the lower knowledge levels of recall and recognition. Often, faculty state that they most strongly desire that their students understand "big picture" ideas and the relationships between central concepts. All too often, however, tests constructed by faculty focus heavily on recognizing the definitions of the boldfaced words in the textbook. Without question, students need to understand the vocabulary of astronomy; however, because boldfaced word questions are easier to write, these low-cognitive-level questions often dominate the tests that students are administered in introductory science courses. When this occurs, students quickly adopt a study mode of memorizing the text rather than integrating ideas and reasoning about complex concepts because doing so leads to higher test scores. One strategy for designing conceptually challenging questions that increase the mental engagement of students is to present standard questions in unusual contexts. For example, a typical question about identifying moon phases can be enhanced by asking about a phase of one of Jupiter's moons as seen from another of Jupiter's moons. (Additional examples are found in Appendix D.) We should remind you that students find higher-level questions difficult, and an overreliance on such questions can result is low test scores and, more important, low class morale; it is important to maintain a balance.

STATISTICAL ANALYSIS

The degree to which a particular item is reliably assessing students is quantifiable in a process known as item analysis. When you take your bubble sheets to be machine graded, two standard statistical tests will often be run without you even noticing. The first of these is *item difficulty*. The item difficulty is a number between 0 and 1.0 that describes the proportion of students who answer the item correctly. You may notice that this statistic should probably be renamed item *easy-ness* because a high value means that most students answered correctly. For example, when an item has an item difficulty of 0.40 it means that 40% of students answered this item correctly. Depending on the exact scenario, this is probably a difficult question or one that is probing a common misconception.

> *Exercise*
> If you are deeply interested in evaluating your tests, ask some students to mark on their tests which questions they know or can figure out and which they don't know or can't figure out. We guarantee that you'll be surprised!

The second statistical test for each item is known as *item discrimination* or *biserial-R*. The item discrimination is a value ranging from –1.0 to +1.0 (although more typically it ranges from 0 to 0.5) and is essentially a correlation statistic. The test compares student responses on that item to how students score on the exam as a whole. A high item discrimination value (often 0.3 or greater) suggests that students who scored high on this exam overall tended to get this particular question correct while students who scored low on this exam tended to get this particular question wrong. In other words, the item discriminates between high-scoring students and low-scoring students. At the other end of the spectrum, an item discrimination score near 0, or possibly even negative, suggests that students who generally scored high on the exam did not do better or worse on this item than the low-scoring students. This is usually interpreted to mean that the better students were overinterpreting the question or that the choices were written such that more than one was likely correct. Item discrimination values near zero usually mean that students are simply guessing, possibly because the concept was not adequately covered in class or the text, and the item doesn't successfully discriminate between strong and weak students. Items that have low or negative discrimination should be significantly revised or even discounted in calculating students' scores (assuming that this does not indicate an error on the answer key, which is the other possibility always worth checking). One caution: If a very high percentage of students gets a particular item correct, it will inevitably have low discrimination because the strong students did no better than the weak ones. This does not mean that the item is flawed.

Taken together, considerations of the physical construction, the conceptual depth, and the statistical analyses of the items are great first steps in improving your test item-writing ability.

ADDITIONAL TIPS

Finally, we'd like to share some tips collectively gathered over the years both by educational researchers and our own experience.

- Students have been conditioned to think that the way to get high scores is to figure out what will be on your tests rather than to learn the material. It is therefore worth extra class time to talk about what your tests are like and what your expectations are.
- It is important is to recognize that testing and grading are anxiety-producing issues for students, and some compassion is warranted.
- Student perceptions of "fairness" are paramount. As a result, when you make mistakes or write poor questions, be honest with your students, and do your best to correct your mistakes publicly. One way to average out your mistakes is to administer many exams, limited of course by your time resources for writing and grading.
- We always summarize the results (i.e., number of As, number of Bs, etc.) after the exam so that students can clearly see where they rate relative to the rest of the class—this eliminates the students that say, "Hey, everybody failed this test," when in fact only a small portion of students did poorly.
- When returning tests, we try to stress the positive outcomes: "Hey, can you believe that 25% of the class got As and six of you actually aced the test?" This again helps send the message that the test wasn't "impossible" and that expectations of success are reasonable.
- When students do have grading issues to discuss, it is almost always better to invite students to your office hours and deal with them individually or have them resubmit their tests with a written explanation of their issue. Students can be particularly aggressive when they first

find out they have received a lower-than-expected grade, and public confrontations are better avoided.

- On our exams, we always put some easy, knowledge/recall questions at the beginning to build student confidence. We then increase test item difficulty by asking multistep reasoning questions, not by making questions longer or by using more vocabulary.
- We post the answer key right after the test so that students can figure out how they did virtually immediately. This takes some pressure off test scoring and posting results right away. It also encourages students to keep copies of their tests, which will ultimately be useful for studying.
- Recent privacy laws make it difficult to post student grades on the wall outside your office. However, students will appreciate it if you can find ways to inform them of their standing in the class, perhaps by always bringing your grade book to class or using a password-protected database system such as *WebCT*™, which might already exist in your institution.
- One of the large-lecture halls we use has aisles only along the walls. In this particular environment, students who complete the exam early cause continuous and major disruptions for other students in the class when they get up to leave. So, after explaining the problem several days before the exam (as well as on the syllabus), we politely ask students to bring a book or some homework from another class so they can stay until the end of the class period when we collect and excuse all students simultaneously.
- Cheating issues are painfully difficult, but you can avoid some common problems by stating specific rules on your syllabus. One of us requires students to bring their student ID with them to the test and to show it when they submit their test. We do not allow students to wear headphones. We ask students who wear baseball caps to turn them around backward so that we can watch for wandering eyes. Frequently circulating around the room also helps to prevent cheating problems. When we do catch students cheating, we follow through with the university policy and notify the dean of students. It is important to do this because it is the only way of establishing patterns of behavior.

> *Helpful Hint*
> Announcing and making multiple forms of an exam with the questions appearing in a different sequence is useful for deterring wandering eyes. Photocopying your exam forms in different colors makes even distribution easier and faster, as does making about 10% more copies than you actually need.

Testing and grading issues are deeply troubling issues for both students and faculty. They rank high on student concerns because students need to show a track record of success in order to graduate from college. These issues are understandably difficult for faculty because testing and grading are time consuming and often put faculty in an uncomfortable position of conflict with students. However, the guiding principles we offer help us to think about tests as an opportunity to help students celebrate what they have learned instead of simply a day for us to figure out what students do not know.

GROUP EXAMS

Way out on the edge of the possibilities you might consider, we have recently been having great success with collaborative group exams. We first attempted group exams for two reasons. The first was that we had been using collaborative group exercises in our course with high success and wanted to continue to capitalize on the benefits of this and demonstrate the value of group work—"test as you teach." The second was that nonscience major students often complain about "test anxiety" and we hoped that using a form of collaborative group tests might alleviate some of this anxiety. The procedure we use is to give students two bubble sheets on which to record their answers. We first have students complete the exam individually and submit their bubble sheets (which takes about half an hour). Then we have students complete the same exam again, but this time working with their collaborative groups (which takes about 15 minutes). We have each student complete her or his own second bubble sheet so that if students disagree with the group's consensus on an answer, they can privately dissent. We then give students the average of the two scores. This allows us to ask more difficult questions on the exam because we know that student scores will be improved by working in groups. This does not drastically help traditionally low-scoring students because if a student scores a 40% individually and an 80% with help from their group, he or she still scores a failing 60%. This has had the unintentional benefit of lowering the number of students who miss exams because students who do not take the exam at the regularly scheduled time are not allowed to take the group portion of their exam (and suddenly that reason for requiring the exam to be rescheduled often just goes away).

> *Excellent Resource*
> *The Hidden Curriculum: Faculty-Made Tests in Science*, Shelia Tobias and Jacqueline Raphael, Plenum Publishing, 1997; ISBN: 0306455803

HANDLING TOUGH QUESTIONS FROM STUDENTS

- *I thought you wanted me to choose answer B because ...* Not only are good multiple-choice items difficult to write, but strong students will often "read too much into" the question in trying to provide the answer they think the professor wants. If students can justify why they selected the answer they did in a way that clearly demonstrates that they understand the concept, we typically go ahead and give them credit. Contrary to most students' opinions, giving credit on a single missed item on a 30-item multiple-choice test won't make much of a difference on the student's final grade, and the goodwill might go a long way.

- *I think this is a trick question ...* When students catch a poorly worded question, it is perfectly acceptable to agree and give all students who answered the question an extra point. One way to help you decide if the question was inappropriate is to look at the item difficulty (how many students answered it correctly) and the item discrimination (how did high-scoring students answer this question compared with low-scoring students). If these values are reasonable, then it is best to deal with students individually rather than as a class as a whole.

- *The test took far too long ...* Students who take ASTRO 101 are highly varied. You should expect a few students to complete the exam in about the same length of time you take to finish it; however, our experience is that you should allow three to four times as long for students to complete it. In other words, if it takes you about 12 minutes to complete the key AND to fill out the bubble sheet, this is about right if you have 50 minutes for your exam period.

- *English is not my first language and I need to use a dictionary ...* Although we have numerous foreign students, we generally do not allow them to use a dictionary because the definitions will often unfairly answer the question. On the other hand, our intention is for students to learn astronomy and that language should not be a barrier, so we encourage them to sit near the front and ask us to explain any words that are unclear.

- *I have to miss the exam because ...* When you teach large enrollment courses there will always be students who miss the exam both for seemingly legitimate and totally unacceptable reasons. We always deal with these students in private. How you handle this is highly individual, but we suggest that you make a plan and then stick to it. One of us always schedules students to take the exam early because we want to be able to discuss the exam openly in class as soon as most of the class has taken the test. On the other hand, the other of us has students take the test late so that the likelihood the exam has been leaked to the entire class is lower. In some circumstances, one of us allows students to use the average of their other exams as their missing exam scores while other instructors let students drop their lowest exam score explicitly to help solve this issue.

- *When will grades be posted ...* Students are highly concerned and anxious about grades, so we do everything in our limited power to post some form of student grades (by whatever code or process the university allows) immediately and frequently. We also encourage students to circle their answers on their test papers, and we post the key immediately after the examination.

- *I need to have extra time ...* The research on test taking shows that, for most students, simply giving students more time to complete an exam does not significantly improve their scores. However, there is a small percentage of students, many of whom have been "certified" as having one of several learning disabilities, who will actually get significantly higher scores if certain allowances are made. We always ask students if they have been working with the student services office on campus and, if so, ask that their advisor or case manager call us so appropriate arrangements can be made.

- *I studied all week for this exam and I still failed ...* We generally do not allow students to retake exams. Some students do not know how to use their study time appropriately. One way to help is to volunteer to go over the test question by question with individual students and ask them to explain their reasoning for why they answered each question the way they did. The resulting discussion, albeit time consuming, will help students understand how to study for and how to answer your test questions and might also improve your test writing skills.

- *What do I need to get on the final exam to earn a B? ...* Although most students do not have a highly enough developed level of metacognitive skills to plan their studying to get a particular grade, students are highly anxious about grades. If you can help students to understand their course standing at any time during the semester, students will greatly appreciate your efforts.

- *Do you grade on the curve? ...* When students say "grade on the curve," they do not mean, "do you use a Gaussian distribution and standard deviations to assign grades" with As being $+2\sigma$ and Bs being $+1\sigma$ and so on. Students really mean do we add points to increase everyone's score or apply a variable scheme for assigning letter grades to numerical scores. We do not curve grades, nor do we add points to particular exams. When grades depend on how students do relative to each other, competition is encouraged, and we want our students to help one another. Only by clearly stating the performance levels we want our students to achieve can they possible rise to meet them. Occasionally, we will lower the requirements to achieve certain letter grades at the very end of the semester if we feel that too few students have earned As and Bs. The difficulty in relying on this strategy is that during the semester students then underestimate how well they are doing and can get discouraged.

- *This test was unfair because* ... Students will sometimes challenge you during regular class time. Be aware that testing and student grades are high-anxiety issues for students, and some compassion is appreciated. Whether or not the student has a valid point, it is always best to ask the student to see you during your office hours so as not to devote precious class time to an issue that might best be handled in private. Making changes to how exams are scored and exceptions on grading should be done with as little fanfare as possible. Treading carefully on students' perceptions of fairness is paramount.

Chapter 8
Alternatives to Multiple-Choice Tests

We frequently hear from students that they are "not good test takers." Despite how tempting it is to argue with students that they must be good test takers to have gotten this far in their education in the first place, we do sympathize with students' distaste for the typical multiple-choice exams they encounter in college. From the students' perspectives, multiple-choice items pigeonhole students into a particular set of answers that are not "in their own words." In such a situation, it is imaginable that a student could fully understand a concept being tested and explain it given a chance, yet still not correctly select the required answer from a set of given choices. More worrisome from the professor's perspective is that students taking multiple-choice tests have a reasonable chance of blindly selecting the correct response using an inaccurate conceptual understanding or, worse yet, no conceptual understanding at all.

Tobias & Raphael (1997) have gone so far as to suggest that student success in college is wholly dependent on mastering the "hidden curriculum" of test taking instead of actually understanding course content. Given all of this, it seems prudent to augment multiple-choice test items with other types of testing strategies whenever possible. In this chapter, we briefly describe strategies for using assessment

Box 8.1 Some Alternatives To Multiple-Choice Testing

- short essay questions
- numerical problems
- homework sets
- term papers
- current events
- portfolio assessment
- performance assessments
- concept maps

approaches such as short essay questions, numerical problems, homework sets, term papers, current events, portfolio assessment, performance assessments, and concept maps.

SHORT ESSAY QUESTIONS

A frequent alternative to using multiple-choice items is to probe student understanding by asking questions that require students to answer by providing a short written narrative of 50 to 100 words. Faculty who use short essay questions believe that they can get better insight into the depth of student understanding than is possible with multiple-choice tests. In some sense, this mimics the historical approach to testing in which all students were tested using extended oral interviews, a process that today is only commonplace at the graduate level.

In gross generality, there are three types of short essay questions. The first asks students to reproduce a "story" or a "process" in essentially the same words as was provided to them in lecture or in their text. A simplistic example might be *List the steps in the thermonuclear fusion reaction that powers the Sun.* These are most commonly graded by listing the key elements needed for a correct answer and assigning points for each element provided. This style of question is designed to measure basic knowledge and recall but can query longer stories and more steps in a process than a multiple-choice question. Questions asking students to provide definitions or provide labeled sketches also fall in this category.

> *Excellent Resource*
> An extensive collection of samples and descriptions of alternative assessment methods is available online in the *National Institute for Science Education Field-Tested Assessment Guide* at URL: http://www.wcer.wisc.edu/ nise/cl1/flag/.

The second type of short essay question is one in which students are prompted, "Explain your reasoning" when analyzing a given scenario. An example might be *Two stars with identical luminosities have very different temperatures. Which star has a larger diameter? Explain your reasoning.* This is probably our favorite type of short essay question because it requires students to apply basic concepts to analyze a complex scenario. This type of question is also useful as an addendum to a multiple-choice item because it provides a window into how students arrived at their response (but it cannot be machine graded). The responses also provide insight into what needs to be revised in your course because it highlights which concepts students are understanding and which they are struggling with.

The third type of short essay question asks students to synthesize several ideas or evaluate scenarios that have not been presented to them in class or in their texts. For example,

You have been given $6000 to spend on telescope equipment for a local outreach program at the community museum. Using the telescope and binocular catalog provided, explain exactly how you would allocate this money and justify each piece of equipment purchased.

This type of question, which might have the initial appearance of being simple, asks students to work at the higher levels of Bloom's Taxonomy of Educational Objectives and asks students to use their knowledge to create and justify decisions as well as evaluate scenarios.[1]

The major disadvantage to short essay questions, and longer essay questions for that matter, is that grading them can be a time-consuming task, which is primarily why they are uncommon in large-lecture courses. Another disadvantage is that students sometimes think that grades on such tasks are somewhat subjective (see our argument in the previous chapter against this truism) and are more apt to ask for reconsideration of grades. Finally, the grading of student writing can be woefully depressing given the abundance of spelling mistakes, poor sentence construction, unimaginably weak responses, and ultimately lower grades than often result from using multiple-choice tests. Certainly most student work does not fall in this category, but enough does that some faculty avoid collecting student writing all together to stave off depression!

Fortunately, strategies do exist that make grading short essay questions more expedient and less open to student complaint. Our favorite approach we call *performance scoring*, partly because our goal is to measure and record student performance on a series of questions and partly because it is done at break-neck speed. We do not correct students' mistakes by providing any written commentary on their tests other than a numerical score. And, although it is a point worth debating, we typically do not consider grammar or spelling unless it is so poor that we are unable to find meaning in what students have written.

---

[1] Faculty sometimes refer to all short essay questions incorrectly as open-ended questions. In fact, only the last type of question is an open-ended question. It is open ended because there are multiple correct responses that students could give, if appropriately justified. In contrast, the first two types of short essay questions are called "student-supplied-response items" because there is only one correct answer.

Box 8.2 Performance Scoring Chart

Score 2: The answer has most of the aspects we are looking for, and there are no gross errors or omissions in the response.

Score 1: The answer has only some of the important aspects we are looking for, or there is a significant error that indicates a partial but significant flaw in the student's mental model.

Score 0: The answer has almost none of the important aspects we are looking for, and there is a significant flaw in the student's response. This score is also earned for no serious attempt at the question provided or if the student responds to a different question, even if correct.

When using this sort of scoring, it is critical to provide students with a thoughtful model answer before returning the student scores. In fact, if you provide students in advance with model responses to example questions, you will find that students' performance improves. Another way to speed up the grading process is to set up the A, B, C, and D grading scale to be more lenient. For example, if we have five questions at 2 points each for a total of 10 possible points, once could set 8 points (80%) as an A, 6 points (60%) as a B, 5 points (50%) as a C, and 4 points (40%) as a D. This seemingly odd approach has two important consequences. First, it allows us to grade quickly, grade harshly, and be less concerned about making a grade-breaking mistake. Second, it helps to avoid the inaccurate, but somehow real, feeling that many students get, that if they earn 1 out of 2 points on a particular question, then they believe they "failed" that question (because 50% is an F).

NUMERICAL PROBLEMS

Closely related to short essay questions are tasks that ask students to make calculations. Used under the banner that astronomy is a quantitative science and therefore astronomy students should be comfortable with mathematics, numerical problems are common in astronomy more because of most faculty's personal history in taking physics courses than astronomy's conceptually tight relationship to physics. For example, students could be asked to *calculate the length of the semimajor axis for an asteroid with an orbital period of 4.6 years.* For this type of question, the expected response is a series of numerical sentences starting with Kepler's third law and ending with a distance result—a distance value that will likely be scored harshly as incorrect if not provided with the correct units. These sorts of questions are quite commonly found as end-of-chapter problems in most textbooks, sometimes realistically composed from ingenious angles and with significant care to ensure that they are plausible.

The most common strategy to score numerical problems is to assign points to different parts of the problem. For example, we might assign 3 points for writing down the correct formula, 2 points for correctly doing the algebra, 2 points for correctly doing a units conversion, and 2 points for getting the correct final result with appropriate units. It is worth noting that such a grading scheme is easy to use, and most faculty find that few students complain about the way their answers are graded. The exception occurs when a student has 2 points subtracted while a classmate only lost 1 point for an identical mistake—and students DO compare!

We should disclose that, personally, we find little value in asking students to memorize formulas and perform calculations without regard for interpreting the results. Certainly, questions like *Calculate how many kilometers to Alpha Centauri at 4.3 ly away* have some merit both in terms of learning how to covert units and to utilize scientific notation, but these pedestrian skills constitute at most a small part of the big picture on which we want our students to focus when learning astronomy.

A more clever approach, and one that we certainly didn't create, is to ask students to create "new" knowledge—that which is not in their textbooks—by asking what happens to a result if we change the independent variables. Consider a question like *How would Earth's rotation and revolution periods change if its distance to the Sun were twice what it is now and how many days long would a year be?* This deeply probes Kepler's third law, whereas a plug-and-chug period/semimajor axis length calculation does not.

Some faculty find "back of the envelope" estimation questions, sometimes called Fermi problems, to be useful. For example, *How much longer does it take a spacecraft to get to Saturn compared to going to Jupiter?* With such a problem, students are not given the relevant data—something that often sends them looking for a formula containing those things—but instead must determine what is needed and then make educated guesses for the required quantities. Indeed, such questions can be more time consuming to grade, but we submit that they are far more interesting for you to grade and for students to complete. There are two warnings about questions of this type. First, students are often uncomfortable with making estimates having learned all of their lives that science is about getting the single "right" answer. Certainly a high-stakes exam is not the place to introduce this kind of question. Second, it is hard to overestimate just how much more difficult this type of question is than a question of comparable complexity in which the data are supplied. When first using Fermi questions, select problems that would be far too simple if presented in the standard format.

HOMEWORK SETS AND TERM PAPERS

In many smaller-enrollment classes, and sometimes even in larger classes, faculty assign and grade homework sets. These assignments range widely from assigning students to submit written answers to the odd-numbered end-of-chapter questions to term papers and presentations on topics of the students' own choosing. Personally, we think that these are a great idea if grading resources are available (including your own time). If for nothing else, homework deadlines keep students somewhat engaged in thinking about astronomy outside of class.

Historically, the biggest problem with assigning homework was that students would not always submit their own work. Just like a file full of multiple-choice tests are passed down from generation to generation among college students, homework assignments over the years often do not change considerably from textbook version to textbook version or from professor to professor. Today, the problem is compounded by the wealth of resources on the Internet. Our students know all too well how to copy a passage from a NASA Web site into a word processing document and then massage it into their own words. It is certainly open to debate the extent to which this is acceptable, but it does raise concern for many faculty.

An effective approach to circumventing this history is to give students problems that are rich in context by focusing on asking students to create "new" knowledge. For example, a question like *How long would it take for a spaceship to travel between Venus and Mars?* is such a question in that students need to make contextual assumptions about the speed of a spaceship, the relative positions of Venus and Mars, and the units in which they wish to work. Two aspects of this particular question are worth highlighting. First, students will not be able to find the exact distance between Venus and Mars in any textbook, which many will find frustrating; we this think is a good thing in some cases (don't laugh, we're serious here). Second, and more important, context-rich questions give students something meaningful about which to engage in academic discourse because different students will come up with different answers due to different assumptions. In this particular example, the classroom discourse can center on whether the planets are in conjunction, opposition, or quadrature, in addition to the range of realistic interplanetary speeds.

Many faculty state that one of their course goals is for students to become life-long learners aware of current events in astronomy. To address this worthwhile goal, another common assignment is to find, summarize, and submit current newspaper stories or articles on astronomy topics several times, or even weekly, throughout the semester. At the risk of advocating a particular publisher, we direct students who are having difficulty finding readable articles to *Astronomy* or *Sky & Telescope* magazines, which are common in most libraries, as well as *Time* and *Newsweek*.

As mentioned briefly, there is considerable debate on the degree that the quality of student writing should be considered when assigning grades in a college science classroom. One school holds that poor writing (especially spelling and grammar given the ubiquity of spelling and grammar checkers) is strictly an indication of lack of care on the part of students; therefore, high expectations will lead to high quality papers. To some extent this is true, although the level of policing required to enforce this is high and if, as a grader, these technical issues distract you, it can be difficult to discover the student's true meaning. The other school holds that poor writing is really a symptom of poor understanding and therefore student writing should be examined more holistically. Further, if a student has deep problems with writing (which is rare), he or she needs to be referred to a writing center to receive the kind of help that a science teacher with a class of 100 cannot possibly be expected to provide. This perspective suggests that although general comments about writing might be warranted, it is unnecessary to "catch" every last error in a student's paper. Instead, it should be graded for the message it contains and, if the writing is so poor as to conceal the message, it should be returned ungraded so that it can be brought up to a college standard and then resubmitted.

PORTFOLIO ASSESSMENT

Portfolio assessment strategies have their roots in creative arts, such as photography, architecture, and writing. The overall goal of the preparation of a portfolio is for the learner to demonstrate and provide evidence that she or he has mastered a given set of learning objectives. To create a high-quality portfolio, students must organize, synthesize, and clearly describe their achievements and effectively communicate what they have learned. The evidence can be presented in a three-ring binder, as a multimedia tour, or as a series of short papers.

More than just thick folders containing student work, portfolios are typically personalized, long-term representations of a student's own efforts and achievements. Whereas multiple-choice tests are designed to determine what the student *doesn't* know, portfolio assessments emphasize what the student *does* know. They are not vastly different in philosophy than semester-long observing logs; portfolio assessments can provide a forum for extended and complex learning activities and observations with the added benefit that much of the responsibility of both learning and assessment can transfer from the faculty member to the student.

A unique aspect of a successful portfolio is that it also contains explicit statements of *self-reflection*. Statements accompanying each item of evidence describe how the student went about mastering the material, why the presented piece of evidence demonstrates mastery, and why mastery of such material is relevant to contexts outside the classroom. Self-reflections make it clear to the reader the processes of integration that have occurred during the learning process. Often, this is achieved with an introductory letter to the reader or as a summary at the end of each section. Such reflections ensure that the student has personally recognized the relevance and level of achievement acquired during creation and presentation of the portfolio. It is this self-reflection that makes a portfolio much more valuable than a folder of student-selected work.

Three variations we have found to be useful in the context of introductory astronomy are as follows:

- *Showcase portfolios*—A showcase portfolio is a limited portfolio in which a student is allowed to present only a few pieces of evidence to demonstrate mastery of learning objectives. Especially useful as a first step in trying portfolios, a showcase portfolio might ask a student to include items that represent (i) their best work; (ii) their most interesting work; (iii) their most improved work; (iv) their most disappointing work; and (v) their favorite work. Items could be homework assignments, examinations, laboratory reports, news clippings, or other creative works. An introductory letter describing why each particular item was included and what it demonstrates makes this type of portfolio especially insightful for the instructor.

- *Checklist portfolios*—A checklist portfolio is composed of a predetermined number of items to be included, similar to how a course syllabus might have a predetermined number of assignments for students to complete. A checklist portfolio gives students the choice of a number of different assignment selections to complete in the course of learning science. For example, instead of assigning exactly 12 sets of problems from the end of each text chapter, students could have the option of replacing four assignments with relevant magazine article reviews or observation reports that clearly demonstrate mastery of a given learning objective. Additionally, class quizzes and tests can become part of the portfolio if that is what is on the checklist of items to be included. A sample checklist might require a portfolio to have eight correctly worked problem sets, two magazine article summaries, two observing session reports, and two corrected examinations in addition to self-reflection paragraphs where the student decides which objectives most closely fit which assignments.

- *Open-format portfolios*—An open-format portfolio generally provides the most insightful view of a student's level of achievement. In an open-format portfolio, students are allowed to submit anything they wish to be considered as evidence for mastery of a given list of astronomy learning objectives. In addition to the traditional items like exams and assignments, students can include reports on museum or planetarium visits, summaries of television documentaries, imaginative homework problems, and other sources from the "real world." Although these portfolios are more difficult for the student to create and for the instructor to score, many students report that they are very proud of the time spent on such a portfolio.

Portfolios can be used successfully in large courses provided there is an infrastructure for students and instructors to utilize. Most importantly, each item in the portfolio needs to be in a similar format; the use of cover sheets, forms, and prescribed notebooks often helps. In this case, students' creativity is somewhat sacrificed for the sake of uniformity. Finally, if graduate teaching assistants are used, each assistant should take responsibility for a particular series of learning goals, thus becoming an expert of sorts and seeing all student submissions. If announced to the students, this helps curtail academic dishonesty and variation in scoring.

Because each portfolio is individualized, student assessment must be compiled by looking at the portfolio's contents relative to the course learning objectives. Each piece of evidence should be graded according to a predetermined scheme. The items can be scored discretely as a 0, 1, 2, or 3 based on the grader's judgment about the student's presentation as related to the stated learning goals. (A larger scale can be used, but the reliability of different faculty giving the student the same score decreases.)

Box 8.3 Porfolio Grading Criteria

Each individual piece of evidence will be graded according to the following scale:

Score 0: No evidence—the evidence is not present, it is not clearly labeled, or there is no rationale or self-reflection.

Score 1: Weak evidence—the evidence is presented is inaccurate, implies misunderstandings, has insufficient rationale or insufficient self-reflection.

Score 2: Adequate evidence—the evidence is presented accurately with no errors nor misunderstandings implied, but the information is dealt with at the literal definition level with no integration across concepts. Opinions presented are not sufficiently supported by referenced facts or facts are presented without clear relevance to opinions or positions.

Score 3: Strong evidence—the evidence is presented accurately and clearly indicates understanding and integration across concepts. Opinions and positions are clearly supported by referenced facts.

The overall portfolio is scored as follows as an indication of the extent to which the portfolio indicates that the student has mastered the 15 course objectives listed elsewhere in the syllabus:

Grade: Rubric:

A : Strong evidence in at least 12 objectives; adequate in other three
B+: Strong evidence in at least 12 objectives; adequate in at least one other
B : Strong evidence in 10 objectives; adequate in all others
C+: Strong evidence in 9 objectives; adequate in others
C : Strong evidence in 9 objectives; adequate in at least one other
D+: Adequate evidence in 12 objectives
D : Adequate evidence in 10 objectives
F : Adequate evidence in less than 10 objectives

Adapted, from Slater, T. F., Ryan. J. M. & Samson, S. L. (1997). "The Impact and Dynamics of Portfolio Assessment and Traditional Assessment in College Physics." *Journal of Research in Science Teaching*, 34(3). pp. 255—271.

Evidence scored as a 0 or a 1 is rather straightforward based on the criteria listed. The most difficult judgment usually lies between awarding a score of 2 and a score of 3. In particular, a score of 2 is awarded if the student has addressed the learning objective correctly and clearly, but only at the literal-descriptive level; there is little explicit integration across concepts or indication of relevance to the student. To be awarded a score of 3, the evidence must clearly indicate that the student understands the objective in an integrated fashion. Such evidence provides the reader deep insight into the complexity of the student's comprehension.

Viewing student portfolios from this perspective drastically changes the emphasis from collections of facts to encompassing concepts. Such a grading procedure also shifts responsibility for demonstrating competence from the instructor to the student. Effectively shifting this responsibility affects comments placed in the portfolio by the grader; comments are directed toward improving the next submission as well as indicating the inadequacies of the current evidence.

PERFORMANCE ASSESSMENT

Although facts and concepts are fundamental in any introductory science course, some faculty value equally procedural knowledge and analysis skills. Student growth in these latter facets proves somewhat difficult to evaluate, particularly with conventional multiple-choice examinations. Performance assessment is a strategy, used in concert with more traditional forms of assessment, designed to provide a more complete picture of student achievement. Performance assessments have been highly validated by English faculty who conduct writing assessments, Olympic judges who score competition divers, and jury panels who evaluate musical performances.

Most performance assessments require students to manipulate equipment to solve a problem or make an analysis. Students are then scored by comparing the performance against a set of written criteria. Students can either complete the task in front of a panel of judges or complete a written response sheet. When used with students of highly varying abilities, performance tasks can take maximum advantage of judging student abilities by using tasks with multiple correct solutions, thus providing insight into a student's level of conceptual and procedural knowledge.

It is critical to have predetermined criteria to evaluate performance. Students should not be scored/graded against their peers, but rather on the predefined criteria, which should be provided before the assessment. Accordingly, the grade book and student feedback reflect levels of competency, not comparative scores.

Box 8.4 The Telescope Performance Task

Your task is to set up and align the 8-inch telescope, find three different sky objects, and accurately describe some aspects of these objects that astronomers consider to be important.

Grading Rubric

Level 3: Student completes all aspects of task quickly and efficiently and is able to answer questions about the equipment used and objects observed beyond what is obvious. The tasks are (a) align telescope mount with north celestial pole; (b) align finder telescope with primary telescope; (c) center on target object; (d) select and focus appropriate eyepiece; (e) provide information about the target beyond the literal descriptive level; and (f) answer questions about the target correctly.

Level 2: Student completes all aspects of task and provides descriptive information about the equipment and objects observed.

Level 1: Student is not able to complete all aspects of task or is not able to provide sufficient information about the equipment used or objects observed.

Level 0: No attempt or meaningful effort obvious.

CONCEPT MAPS

Professor Michael Zeilik at the University of New Mexico has had considerable success in introductory astronomy using a strategy collectively known as concept maps (Zeilik et al., 1997). A concept map is a diagram of nodes, each containing concept labels enclosed in a box or an oval, which are linked with directional lines, also labeled. The concept nodes are arranged in hierarchical levels that move from general to specific concepts. The core element of a concept map is a *proposition*, which consists of two or more concepts connected by a labeled link. In a concept map, propositions are connected to form a branching structure that represents the organization of astronomy knowledge in long-term memory. The basic assumption of the concept map is that interrelatedness is an essential property of knowledge, and that understanding can be represented through a rich set of relations among important concepts in a discipline.

What underlies this approach is that meaningful human learning, as opposed to rote memorization, occurs when new knowledge is consciously and purposively linked to an existing framework of prior knowledge in a nonarbitrary, substantive fashion. In rote memorization, new concepts are added to the learner's framework in an arbitrary and verbatim way, producing a weak and unstable structure that quickly fades after the students have taken the exam. Concept maps, used in the context of instruction, help students develop more robust conceptual structures. Used in assessment, concept maps are evaluated to determine the extent to which students understand the interrelatedness of concepts in astronomy.

There are several approaches to using concept maps that work in the context of teaching as well as assessment.

- *Fill-in concept mapping*—The instructor constructs a concept map and then removes all of the concept labels but retains the links. Students are tasked to replace the labels in a way that makes structural sense.

- *Select and fill-in concept mapping*—The instructor constructs a concept map and then removes about one-third of the concepts from the nodes. These deleted concepts can be placed in a list for students to choose from. Scoring can be as simple as computing the percent correct; faculty in large classes can use computer-gradable bubble sheets by labeling the words with letters and the nodes with numbers.

- *Selected terms concept mapping*—From a given list of 10 to 20 concept labels, students are tasked to construct their own maps using only these labels. The focus is on students explicating the linking relationships.

- *Seeded terms concept mapping*—Students are given only a small subset of concept labels and tasked to construct a complete concept map using these, and an equal number of additional labels drawn from their own knowledge of the topic.

- *Guided choice concept mapping*—Students are provided with an abundance of concept labels from which students select about half to construct their own maps. When done over a period of time, the instructor can focus on which concepts appear and which disappear.

Because of the creative nature of concept maps, student frustration levels can be very high when concept mapping is first introduced, especially in large classes. To mitigate some of this anxiety and to encourage students to reflect on their own thinking, groups of three or four students can work collaboratively on a concept map, which initially may or may not be related to astronomy. This approach can engender a rich learning experience as peers argue, debate, and cajole each other. The result is a genuine effort at negotiating the meaning of scientific concepts and attempting to reach consensus. As with most collaborative group activities, the power of the process resides in the social construction of knowledge.

Box 8.5 Examples of Concept Maps

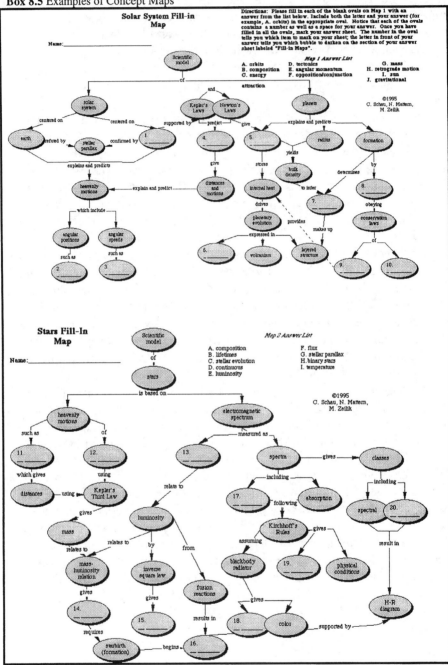

Chapter 9
Course Evaluations: Finding Out What's Working...and What Isn't

One of the most unnerving things you can do as a teacher is to ask the students in your class to take out a piece of paper and list on one side the aspects of the course that are working well and, on the other side, list the aspects of the course that they wish could be changed. Whenever we do this—and we do some form of course assessment frequently in all of the classes we teach—we always feel a sense of uneasiness that maybe we will learn some things about how our students feel about us and our approach to teaching that we would almost rather not know. Ultimately, however, this process never fails to reinforce that most of what we are doing is positive— "They like me...they really like me"—while pointing to areas in which real improvements can be made. Additionally, most students appreciate that asking for feedback is an act of faith and trust, which contributes to the sense of an educational partnership. No single act in the classroom sends the important message of caring more than a professor opening up to, and acting on, honest feedback from students.

Of course, this kind of assessment only provides a glimpse of the entire picture. It provides a general sense of how much students are enjoying the class and can give some insightful guidance as to which course elements students feel are most beneficial in helping them learn. What it does not provide is detailed information on how well the students are really learning. A colleague once proclaimed to us that he could judge what his class understands just by "looking into the students' eyes." There is no doubt that this extremely popular and award-winning professor truly believed this to be true but, as described in Chapter 3, research in physics and astronomy education, bolstered by our own personal experiences, provides a credible challenge to such intuitive claims. In fact, it is our position that if you do nothing to measure student understanding in a regular and systematic way through the course of teaching any topic, it is likely that major areas of student misunderstanding will go undetected until revealing themselves on the final exam, and possibly not even then. Similarly, if your overarching course goals include enhancing student attitudes about science or engendering life-long interest in astronomy beyond your class, you should acquire specific assessment data for this as well.

Any process to gather data during a course for the purpose of improving learning—whether data about students' attitudes or content understanding—is called *formative assessment*. This is in contrast to *summative assessment*, which is data gathering for the purpose of making an evaluative final judgment such as assigning final grades. As a general rule, it is difficult to get do both formative and summative assessment at the same time because students will not usually provide meaningful feedback to you if they think that their grade might depend in part on what they say. However, a few assessments can serve both purposes depending on how the data are used. For instance, a final exam is normally used to determine students' standing in the class, but a careful analysis of the results (see the description of statistical item analysis in Chapter 7) can reveal persistent student difficulties and suggest changes for the next semester. However, in this discussion, we limit formative assessment to only ungraded, usually anonymous, assessments that have no implications for students' grades. And, to simplify the discussion, we divide formative assessment into two broad categories: course assessment and learning assessment.

COURSE ASSESSMENT

An oft-heard complaint from students is that they devote considerable time each semester to filling out course evaluation forms and yet nothing seems to change—lousy faculty keep getting assigned to courses. It is our perspective that the results of these student-completed evaluations, as well as other means of assessing teaching effectiveness (see Chapter 9), are becoming increasingly—and appropriately—important. But, even if changes do result, it is too late for the students in that particular class to benefit. Although we can assure students that future generations will benefit from thoughtful responses, this is often of little consolation. We firmly believe that the single most important thing we have done to help our teaching is to begin honestly soliciting student feedback early in our courses and responding to it. Note that we say "respond" and not simply, "acquiesce." We are not saying that we implement every change our students suggest (*No, I will not supply pizza every Friday!*) but rather that we consider all of their comments carefully and, only where appropriate, make midcourse corrections to address their needs.

As with all robust assessments, there are three important steps in conducting meaningful early-semester course evaluation: data gathering, data interpretation, and data reporting (i.e., closing the loop).

Data Gathering

Possible data gathering methods range from the very informal, like that described in the opening paragraph (write down what you like and what you don't like), to the very formal, such as structured focus-group interviews, described later. Each approach has its strengths and weaknesses, but for those who have never tried to look critically at their own class, the method that you choose is not nearly as important as that fact that you are doing something. Although not exhaustive, the following list provides a range of possibilities that could be easily adapted to fit your needs.

- *Informal written response*—This is the method already described in which you ask students to write their answers to a few simple questions, such as, "What part of this class is most important in helping you learn and what part is least important? Please explain." If you have introduced something new to the course, you could ask directly how they feel about it: "Do you think that we should continue using class time for collaborative group activities? Please explain." We do recommend that you use questions designed to have students reflect on both the positive and negative aspects even though it is mostly the negative aspects that are of interest when thinking about possible course corrections.

- *Check-box questionnaire*—A questionnaire of your own design is a quick, simple, and effective way to generate quantitative data on a wide range of topics. In addition to measuring students' attitudes toward various aspects of the course, questionnaires can also be used to learn about your students by asking demographic questions such as age, major, or hours of work each week. In collecting student attitude data, questions are normally phrased so that students respond on a 5-point scale, called a Likert scale. Possible ways to set up this scale are shown in Box 9.1.

Box 9.1 Options for 5-point Likert scales

Please rate the effectiveness of each of the following course elements in helping you learn astronomy. Use a scale from highly effective to completely ineffective
A. Lecture:
   Highly Effective [ ]   [ ]   [ ]   [ ]   [ ] Completely Ineffective
B. Homework Assignments:
   Highly Effective [ ]   [ ]   [ ]   [ ]   [ ] Completely Ineffective
C. Reading Assignments:
   Highly Effective [ ]   [ ]   [ ]   [ ]   [ ] Completely Ineffective

Please rate your agreement with each of the following statements using a scale of 1 to 5, with 1 meaning strongly agree and 5 meaning strongly disagree.
D. I feel that that availability of the course notes on the class Web site encourages me to skip class more than if the notes were not available.
   [1]   [2]   [3]   [4]   [5]
E. I think that astronomy is easier for some people to learn because they have a natural ability to think scientifically.
   [1]   [2]   [3]   [4]   [5]
F. I will likely get a poor grade in astronomy because I am not good at mathematics.
   [1]   [2]   [3]   [4]   [5]

Please rate your interest in the following topics for discussion at the end of the course.
G. Search for life on other planets
   [ ] none   [ ] little   [ ] moderate   [ ] high   [ ] extreme
H. Latest results from the Hubble Space Telescope
   [ ] none   [ ] little   [ ] moderate   [ ] high   [ ] extreme
I. The scientific accuracy of *Star Trek*
   [ ] none   [ ] little   [ ] moderate   [ ] high   [ ] extreme

Note that these scales are easily adapted for use on machine-readable bubble sheets, which greatly simplifies data analysis. We have shown scales with just the endpoints labeled (questions 1 through 6) and with all options labeled (questions 7 through 9). There is no compelling reason to choose one over the other so, based on simplicity, we generally advocate only labeling the endpoints of the scales. This sort of quantitative data can be entered easily into a spreadsheet and average values and standard deviations for each question can be tabulated quickly.

- *Group interview by colleague*—In this technique, you invite a colleague to visit with your class, typically for about 20 minutes, while you leave the room. You assure the students that you have invited the colleague to do this (otherwise, it can be perceived as some form of administrative review, causing students, in their desire to protect you, believe it or not, to be less forthcoming) and that their identities will be protected (which is particularly easy if your colleague doesn't know your students). We advise that you seek help from someone you trust and, preferably, from someone with experience. We also suggest that you have a brief discussion with your colleague before the interview so that he or she knows the kinds of issues that you are concerned about. Acting as interviewers within and beyond astronomy, we have had success using both highly structured and almost completely unstructured approaches (see Box 9.2).

Box 9.2 Group Interview Approaches

Opening

Hello. My name is _____ and, as your instructor explained, I am here at her request to help her learn something about your perspectives on how ASTRO 101 is going this semester. This is not part of any formal review. After our conversation, I will summarize your comments and discuss them with your instructor and no one else. If there are any comments that I feel could be potentially damaging to the class morale, I will save those until the end of the semester. I encourage you to speak freely and openly, and I guarantee that I will not do anything to identify you individually in my report.

Highly Structured

1. *How well do you feel that your previous courses have prepared you for this course?*
2. *How does the amount of time required out of class in this course compare with other courses you are taking this semester?*
3. *What has been the most difficult topic you have studied so far this semester?*
4. *[Continue with similar questions.]*

Unstructured

What do you think your instructor needs to know about how this course is going so far?

Independent of the level of structure, it is important for the interviewer to ask follow-up questions and ask for examples to keep the discussion active. Even more critical, the interviewer must constantly monitor who is responding to ensure that the opinions of a small but vocal minority do not dominate. We have found it helpful in this regard to poll the class informally by reflecting responses back to the class ("What I am hearing you say is…"), asking for a show of hands from those in agreement, and asking for any other views. Although the interview can be tape recorded, we have found hand written notes to be sufficient.

■ *Focus group interviews*—Focus groups are widely used in marketing research and in designing political campaigns. They normally involve a trained facilitator and a group of 6 to 10 participants; the interviews are normally audio- or videotaped for later analysis. Compared with individual interviews, focus group interviews have the advantages of being more efficient and of generating discussion as participants react to each other's comments. Actually, it is this student-to-student interaction that is the primary advantage of focus group interviews compared with large group discussions, which tend to be a series of one-on-one conversations with the interviewer rather than a true discussion among participants. Although we have never had access to a cadre of trained focus group facilitators, we have nonetheless found it useful to break up our class into groups of 10 to 15 and send each group off with colleagues to a prearranged room for a short interview. Compared with the whole-class interview, this is more resource intensive, but generally the quality of the data is better both because of the smaller size of the groups and because the data are filtered through a broader range of interviewers. If you choose to record the interviews, it is especially important to tell the students that you will not listen to the tapes until the following semester (and follow through on this promise).

- *Observation by a critical friend*—Although this form of data collection does not involve students, using a critical friend to act as a confidential peer observer and provide you with a list of strengths and weaknesses will go a long way in helping you improve your teaching. If a confidential peer is not available, then watching a videotape of yourself will highlight a long list of things you didn't even know you did.

Data Interpretation

Having gathered the data, the next important question is, What are you going to do with it? The type of data you have gathered—qualitative or quantitative—will, in large part, determine the answer to this.

Qualitative data, like that gathered from student writing or interviews, have the advantage of providing depth. The biggest risk is that your overall assessment can be influenced too easily by a minority opinion—especially one that is particularly negative. (Who among us has not lost sleep over that one particularly negative course evaluation from a student even when many more were equally positive?)

It is for this reason that, in the case of analyzing written response data, simply reading the responses is not enough. Even if not overly detailed, we encourage some elementary organization of the data to look for trends. The simplest way to do this is to use different colored highlighter markers to indicate generally positive and negative comments worth reviewing (and scissors if both appear on the same page). Once you have read through all of the comments, or a representative sample in the case of a particularly large class, begin making piles of related comments based on a few broad categories (which you may alter during the process). Finally, review the comments within each category and write some brief notes in which you record not only the theme but also its prevalence. Remember, not every negative comment will require action—12 students claiming that the course is too difficult is not alarming if another 18 report that it is one of their easiest classes! You will be surprised at how little time this process actually requires using a simple tally sheet or a computer spreadsheet.

The other forms of qualitative data we have discussed—large group and focus group interviews—rely on the data collectors for analysis. At the very least, you should plan to sit down with the interviewer(s) to gain a general impression of what was learned and any recommendations for change. Beyond this, we would urge you to request a written evaluation of the interview with a list of recommendations both because it encourages a more thoughtful review of the data and because you might decide that the report can become part of your teaching portfolio (see Chapter 9). Professional interview transcription and analysis can take as much as 30 hours for each single hour of audiotape and is generally not warranted for this personal level of assessment. However, if interviews were recorded, we do urge that the tapes be reviewed the following semester—to protect students and provide some distance for you.

As scientists, we are all generally comfortable dealing with quantitative data, which we know how to manipulate and display. Looking only at averages can sometimes be misleading and so we recommend that, for each question, you compute the frequency of each response in addition to the average and then create a means of displaying the data summary in graphs or tables for the class.

Independent of the type of data, the analysis must include your personal assessment of what you have learned and what you might do about it—something we urge you to write down. Areas of difficulty in your courses uncovered through assessment can be typically divided broadly into four categories:

1. *Simple fixes*—Sometimes, you discover that there are small procedural things you are doing that could be changed easily to match student wishes. For instance, you might learn that the top 12 inches of your whiteboard can't be seen from the back row, the solution to which is obvious.

2. *Fixes for next semester*—It is not unusual to uncover something that would have been changed easily in the syllabus had you anticipated it but cannot be changed during the semester. This might be a simple technical matter like the timing of your office hours (which can be difficult to change with all of your other commitments) or a policy issue like the apportioning of grades for homework (which you might agree is inappropriate but you shouldn't change from what is on your syllabus during the term).

3. *Offsetting concerns*—One of the truisms of teaching is that there are no "one size fits all" solutions. Sometimes, for every student who complains about a particular issue, another applauds it. Although this usually means that there is no need for you to change, you should tell

the class of the results so that they are aware that their issue has been identified but, contrary what students tend to think, their views are not held universally.

4. *Issues without a remedy*—It is sometimes the case that there are important issues, ones with which you sympathize, for which no solution is available. We never take this position lightly—sometimes, creative solutions can be found—but, occasionally, difficult issues simply must be acknowledged to the class and an explanation offered.

It is sometimes the case that there is no appropriate response to any of the issues raised that can be implemented immediately. However, we suggest that having something tangible to take back to the class as an immediate and genuine response to the effort they put into completing the assessment is critically important and, with a little creative effort, something substantive can usually be found. Neglecting to close the loop will result in students being unwilling to expend the effort required to provide meaningful feedback in the future.

Closing the Loop

It is a mantra of the assessment movement that assessment data, to have any value at all, must inform decision making and that those decisions, including the rationale for them, must be made public. The loop is closed when an action is taken and the assessment cycle begins again.

Although this is the shortest section of this discussion, it is unquestionably the most important. As already stated, students are already asked to provide a fair amount of feedback on end-of-semester evaluation forms and, as far as they are concerned, this information all too often disappears into a black hole. To involve students as partners in this process, it is imperative to report the results of the assessment and discuss what changes, if any, will be implemented in response. When changes cannot be made to address a particular issue, a rationale must be provided.

LEARNING ASSESSMENT

Course assessment in the service of improvement as just described remains far from being standard practice within most college classrooms. In contrast, the use of homework, quizzes, and tests to find out how much our students have learned is something that we all do regularly. Moreover, even in less formal settings such as office hours, we all ask questions of struggling students because we recognize the need to know more about a student's specific difficulties if we want to assist him or her in learning. But it is our perspective—and we are certainly not the first ones to suggest this—that such activities only begin to scratch the surface of what can be learned through concerted and targeted efforts to assess student understanding. The use of techniques designed specifically for assessing student learning, as opposed to assigning grades, can have profound effects on teaching effectiveness.

Improvement results both from instructors gaining an improved understanding of areas of student difficulty and from students coming to appreciate better their own areas of weakness (before discovering them on the test).

> **Excellent Resource**
> *Classroom Assessment Techniques* (2nd edition), Angelo and Cross, Jossey-Bass, 1993; ISBN: 1555425003
> *This "bible" of classroom assessment presents many classroom-tested techniques as well as a more comprehensive description than we provide here.*

It is most important that the techniques used for assessing student difficulties be quick, easily administered, low risk for students, and easily analyzed. Further, the assessment strategies need to be used in such a way as to provide either immediate feedback to students or results that can be shared with students at the next class meeting. Although not comprehensive, the following represents some of the techniques that we have found useful and might provide a departure point from which you can develop others.

- *In-class questioning*—The questioning techniques described in Chapter 5, including think-pair-share and talk to your neighbor, are forms of classroom learning assessment that provide immediate feedback to you and the students.

- *Muddiest point*—Hand out 3½ × 5 index cards at the start of class. Near the end of class, give the students a couple of minutes to write down what they consider to be the most confusing point from that day's lecture. Collect the cards as they are leaving and, immediately after class, review the responses (or a sample), looking for trends. If there are one or two consistent points, this makes a great place to begin next class.

- *Ungraded quiz*—Students often complain that they study for tests but don't know what to expect. You can do a lot to alleviate their concerns by having them complete ungraded quizzes (ideally using questions from old tests you have written). These quizzes can require either written response or be in multiple-choice format, and the answers can be discussed immediately after they are collected. (Even if students are told that they will not be graded, there is still a desire among many students to put their best foot forward and they might change their answers before submitting if the correct answers are discussed first.)

You can quickly review the quizzes, looking for areas of difficulty to address at the next class.

- *Weekly reports*—Some faculty have had considerable success by asking students to submit a one-page description of the most important points covered in class each week during the course. Such an assignment encourages students to synthesize the concepts frequently and provides faculty with insight about what their students believe is most important. Papers written by students each week usually answer specific questions such as "What did you learn this week? What questions remain unclear? What questions would you ask your students, if you were the professor, to find out if your students were understanding the material?"

- *Self-assessment of learning gains*—By comparing students' self-report of knowledge (i.e., "How well do you understand...?") with their answers on quizzes, we have found that students can successfully report which concepts they understand and which they do not. Using a quick survey with items like those in Box 9.3 provides you with a glimpse into where students are feeling low levels of confidence. Surveys like these can be administered confidentially via the Internet as well.[1]

Box 9.3 Example Self-Assessment Question

Rate your understanding of each of the following using a scale from 1 (no understanding) to 5 (complete understanding):   A)   Implications of Galileo's observations of the phases of Venus   [1]   [2]   [3]   [4]   [5]   B)   Seasons on Mars   [1]   [2]   [3]   [4]   [5]   C)   The causes of type I and type II supernovae   [1]   [2]   [3]   [4]   [5]

- *National conceptual diagnostics tests*—In astronomy, the *Astronomy Diagnostics Test* (ADT),[2] which is included in Appendix C, has established a national database of thousands of students at all different types of institutions in different regions with pretest and posttest scores that you can use to see how your teaching, and your students, stack up nationally.

---

[1] Fellows of the *National Institute for Science Education* have created a free and easy-to-use Web interface for faculty to administer and analyze self-report surveys at URL: http://www.wcer.wisc.edu/salgains/instructor/.

[2] The *Astronomy Diagnostics Test* (ADT) and its national database of comparison scores is available online at URL: http://solar.physics.montana.edu/aae/adt/.

As a final note, we should mention that examinations in and of themselves can be used to provide a wealth of information to inform teaching practices and not just to assign grades. As described in Chapter 7, item analysis on multiple-choice exams can be used to identify areas of difficulty as well as discover particularly good questions for use during class in, for instance, a think-pair-share exercise. If you are using short essay questions on exams, we suggest making brief notes of recurring problems when you are grading and that, if you have teaching assistants grading your exams, you ask them to do the same. Knowing that a particularly important concept was not understood at the level you expected can suggest changes to the way that you teach this topic or, as sometimes happens, a recognition that a deep understanding of the topic in the time you have allocated is beyond the majority of the class and thus requires changes to your learning objectives related to that topic.

Chapter 10
The Teaching Portfolio: Demonstrating Excellence in Teaching

It's no secret that colleges and universities are placing a greater and greater emphasis on high-quality teaching. Such an emphasis is probably most evident when manifested in the tenure or promotion process—although the requirements are often vague and admittedly play a lesser role than assessment of research quality. Beyond this, demonstrating excellence in teaching can play a positive role in annual reviews for pay raises and in giving you the edge to be on the more interesting university committees while avoiding the less desirable assignments. Moreover, given the commitment of both time and intellectual resources invested in high-quality teaching, there is an intrinsic desire to have this reflected in performance evaluations. Finally, the process of documenting your teaching effectiveness should be one of thoughtful reflection that can, in itself, lead to better teaching.

As with documenting excellence in research endeavors, documenting teaching effectiveness is fundamentally a matter of collecting and presenting evidence. As far as research goes, most of us received the sage advice that we should keep a box in our office that is added to every time we have a paper published, a grant submitted, a newspaper article written about our work, or when we receive letters acknowledging our students' awards in research. These natural artifacts of a robust research program are mostly natural by-products of the process—they don't require us to do much beyond maintaining an active research program. But what

> **Early in any new faculty position, it is imperative that you become familiar with your departmental, college, and university requirements for documenting teaching effectiveness.**

about the evidence we submit for excellence in teaching? Good teaching does not generate artifacts quite so naturally as good research does; even if you keep a teaching box in your office, what should you be putting in it in preparation your performance review?

In attempting to answer this question for yourself, the most important advice we can give is to become familiar with your departmental, college and university promotion and tenure (P&T) requirements at the onset. These sometimes make specific demands on the kinds of evidence you can collect. We further suggest that you need to look far beyond the rhetoric by asking faculty who have sat on many P&T committees recently and become aware of how decisions are actually made. Our position, however, is that this should be viewed as a minimum requirement—often one that just requires that you keep a list of courses, enrollments, and student evaluation scores—and that your case for the quality of your teaching can be greatly enhanced by adding additional material in the form of a teaching portfolio.

A teaching portfolio is a collection of evidence designed specifically to provide reviewers with an in-depth understanding of your scholarly approach to teaching. As such, it is intended to do much more than just show that you are organized, know your subject, and that your students like you. A teaching portfolio not only provides multiple external measures of your teaching effectiveness (including those just listed) but also features your thoughtful reflections on what the evidence says about your teaching and what actions you have taken and will be taking in response to that evidence. A teaching portfolio highlights the strengths and weaknesses in your teaching with an emphasis on your growth and evolution.

Although its usefulness for evaluation has increased the interest in the teaching portfolio, we should emphasize that the very maintenance of a teaching portfolio often leads to teaching improvement because of the reflective nature of the process. To realize the benefits of a portfolio both

> **Demonstrating excellence in teaching often requires letters of recommendation from students. The best advice we've received is to keep a file of students' names with whom you have developed relationships during your classes so that they can provide letters when it comes time for your teaching review. Add to this at least once a semester.**

in providing a structure for personal reflection leading to improvement and in effectively demonstrating growth and reflective practice for the purpose of performance review, your teaching portfolio should be started as early in your teaching career as possible.

The purpose of this brief chapter is to outline one possible structure for a scholarly teaching portfolio that, in the absence of detailed guidance from your institution, you could use to document your growth and success in the classroom. We also provide some suggestions on how the portfolio could be evaluated. Your teaching portfolio, in combination with the outside reviews, comprises your *scholarly teaching* documents. Because every institution and department is different, we suggest that you critically evaluate this model and adapt it to fit your own situation. If your institution does have specific guidance related to documenting teaching effectiveness, you may still find elements in this model that you could provide in addition to those required.

CONTENTS OF YOUR TEACHING PORTFOLIO

In this section we list all of the elements you should consider collecting for your teaching portfolio. We consider the items indicated with an asterisk (*) to be essential.

The materials presented should represent thoughtful and reflective teaching. There is no expectation that every course design, instructional approach, or student assessment strategy be flawless in its implementation. Accordingly, the materials presented as evidence will be most informative if they demonstrate growth. The key to demonstrating this growth and improvement as a teacher resides in the accompanying candidate-written explanations. These explanations should describe succinctly how and why courses were designed and structured the way they were; the specific goals of each course; how the instruction attempted to achieve these goals; how the student assessment approaches supported these goals; and what evidence is available that shows that these course goals were met. Evidence of course revisions based on candidate-collected data will be highly valued by most expert reviewers. In short, although this process is known as an in-depth assessment of teaching, a focus on student learning often makes the strongest case for effective and excellent teaching.

1. *Statement*—A brief (up to 500 words) statement in which you describe your approach to teaching and learning. You should specifically address how you gauge the level of student learning. We recommend writing this statement right away and then reviewing and updating it

annually. This provides a good way to evaluate the growth in your views of teaching and learning.

2. *Course list*—List the courses you taught during the review period, number of credit and/or contact hours for each course, and number of students per course. It is helpful to update this list annually. It is also recommended that at the time of review the Department Head supply comparative information to help reviewers interpret the teaching load within the department.

3. *Student evaluation of faculty forms*—Most institutions require the use of some type of student evaluation form, and you should keep the forms and the statistical reports in a safe place. At the time of review, your Department Head or designate should summarize the results, including a brief synopsis of written comments. In addition, we encourage you to supply a brief narrative offering your interpretation of the results. For instance, just as in the lab, not all experiments are a success, and if you feel that the introduction of something new into your class led to lower-than-average scores, you should explain this. Other forms of student feedback (e.g., focus group interviews) can also be included in this section. Because the reviewers will not necessarily be acquainted with your institution's particular campus culture or norms (if external reviewers are used), we recommend that the Department Head supply information to aid in establishing the context of the numerical data. This could include, for instance, departmental and/or college averages (where appropriate) either collectively or disaggregated by course level (i.e., freshmen, sophomore, junior, senior) or course type (i.e., survey, major, nonmajor, elective, required, large-lecture, laboratory).

4. *Course materials*—For each of a minimum of two different courses that you taught, supply the course syllabus, a list of course goals, a sample student assignment, a sample examination, and any other relevant course materials. For each course you should then supply a brief narrative explaining why the course is designed the way it is, how it coordinates with other courses or programs, and how the evidence presented is designed to help students meet the course goals. We suggest compiling this regularly by adding materials from new courses or newly designed courses.

5. *Student work samples*—Where appropriate, you can supply student work samples as evidence of improvements in student understanding or

performance. Examples that demonstrate student growth are more useful than exemplary final products, and we caution against focusing on the work of only your top students. An interpretative narrative describing how your teaching influenced the work should accompany each work sample.

6. *Video*—Consider including a 10- to 30-minute video clip that demonstrates your classroom teaching with your written description explaining the context of the video clip, the learning goals addressed during the segment, and why it exemplifies your teaching abilities. We recommend arranging to have several lessons videotaped in every course you teach and creating a labeled library of teaching videos. Although you may include only one video in your performance review, we suggest that you review all tapes and make a brief written record of your responses.

7. *Classroom observations*—We recommend that having colleagues observe and report on your teaching should be a regular part of your practice. Reviews by multiple observers should be included for a variety of the courses you teach. If your department does not have specific policies and procedures for conducting classroom observations, we suggest working with your colleagues or Department Head to develop a protocol that includes specific instructions on how to conduct and report classroom observations.

8. *Letters*—As part of the review of your teaching, letters from students describing their experiences in your courses should be collected. Although you may collect letters on your own (or save unsolicited notes that your receive that reflect well on your teaching) the letters solicited by the department will carry more weight with reviewers. The number of letters and the process for collecting them is often dictated by university policy. The one thing you may be able to do is provide a list of names of students from whom you would like letters solicited (normally in addition to letters solicited from students selected randomly from class lists). Because it can be difficult to remember the names of students several years later, we recommend that at the end of each semester you list the names of a few students from each class that you feel you have been especially effective in reaching.

9. *Evidence of innovation*—You should provide evidence of any innovations and an explanation for why the evidence demonstrates

innovation in teaching. Assessment data on the effectiveness of the innovations can be particularly important in making the case for your teaching effectiveness. This is something that it best done at the end of each semester when the experience is still fresh.

10. *Contributions beyond your classroom*—If you are involved in educational efforts that extend beyond your classroom, you should include evidence of this. This could include such activities as textbook writing, K-12 curriculum development, involvement in professional societies, or writing about teaching innovations. In cases where these activities have direct impact on your classroom, they should be included in *Section 9: Evidence of Innovation*. Otherwise such materials may be included in this section, which we recommend be reviewed separately by the external reviewers. We encourage you to supply a brief written interpretation of the materials.

The reason that this section is distinct from general innovations is that the higher-education literature suggests that the link between writing about teaching and teaching effectiveness is weak at best. However, in cases where these efforts cannot appropriately be included within the research section of your portfolio, they should be included here.

The maintenance of a teaching portfolio should not be an onerous task. By making it part of your regular teaching routine to collect this information and write about your experiences, you will be well positioned at the time of P&T application and, more important, you will find that the process itself leads to better and more reflective teaching.

Appendix A: Seasonal Stars Lecture Tutorial

PART I: MONTHLY DIFFERENCES

Figure 1 shows a heliocentric, perspective view of the Earth-Sun system indicating the direction of both the daily rotation of the Earth about its own axis and its annual orbit about the Sun. You are the observer shown in Figure 1, located in the northern hemisphere. You are facing toward the southern horizon.

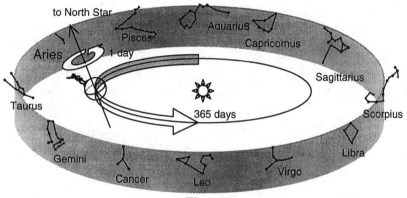

Figure 1

1. Which labeled constellation do you see highest in the southern sky?

2. What constellation is just to the left (i.e., east) and what constellation is just to the right (i.e., west) of the highest constellation at this instant?
   left: right:

3. Noting that you are exactly on the opposite side of Earth from the Sun, what time is it?

4. One month later the Earth will have moved one-twelfth of the way around the Sun. You are again facing south while observing at midnight. Which constellation will now be highest in the southern sky?

5. Do you have to look east or west of the highest constellation to see that one that was highest one month ago?

6. Does the constellation that was highest in the sky at midnight a month ago now rise earlier or later than it rose last month? Explain your reasoning.

PART II: DAILY DIFFERENCES

Figure 2 shows the same Earth-Sun view, including the bright star Betelguese, which is between Taurus and Gemini.

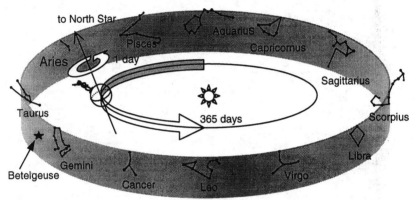

Figure 2

7. Last night you saw the star Betelguese exactly on your eastern horizon at 5:47 P.M. At 5:47 tonight, will Betelguese be above, below, or exactly on your eastern horizon?

Two students are discussing their answers to question 7.

Student 1: *The Earth makes one complete rotation about its axis each day so Betelguese will rise at the same time every night. It will therefore be exactly on the eastern horizon.*

Student 2: *No. The constellation Taurus rises earlier each month and so it must rise a little bit earlier each night. Betelguese must do the same thing. Tonight it would rise a little before 5:47 and be above the eastern by 5:47. You are confusing the sidereal and solar day.*

8. Do you agree with Student 1 or Student 2? Why?

9. How long did it take for Betelguese to return to exactly the same position it was on the previous night—slightly less than 24 hours, exactly 24 hours, or slightly more than 24 hours?

10. The celestial sphere in Figure 3 helps us picture the motion of the night sky from a stationary Earth by imagining the sphere to rotate approximately once every day. To make this model consistent with your answer to question 9, how much time should the celestial sphere take to complete one revolution about a fixed Earth—slightly less than 24 hours, exactly 24 hours, or slightly more than 24 hours? Check your answer with a nearby group.

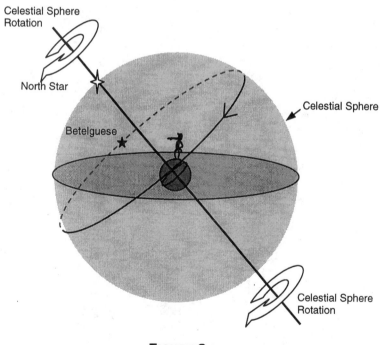

FIGURE 3

Appendix B: Sample Think-Pair-Share Questions

The following list of questions has been selected primarily for use in a think-pair-share format during classroom instruction (see Chapter 5). Questions that are useful for this focus on persistent student difficulties and lead to productive discussion between students. Therefore, such questions are not always "fair" for use in testing situations in which grades will be determined. We encourage you to try these questions, adapt them for your own use and your own terminology, and start creating your own bank of instructional questions that match the content on which you choose to focus. All questions are written for a northern hemisphere perspective but can be easily adjusted.

ANCIENT ASTRONOMY AND THE CELESTIAL SPHERE MODEL

1) Which of the following did <u>not</u> provide the ancients with evidence of a spherical Earth?
   a. Boats disappear over the horizon.
   b. **The Moon casts a circular shadow on the Earth.**
   c. Which stars are visible changes as you go south.
   d. These are all evidence of a spherical Earth.

2) Knowing that the Sun and Moon "look" about the same size in the sky (i.e., they subtend the same angle), what other piece of information do we need to determine the relative sizes of the Sun and Moon?
   a. the size of Earth's shadow on the Moon
   b. the relative sizes of Earth and the Moon
   c. **the number of times farther away the Sun is than the Moon**
   d. the size that Earth would appear if viewed from the Moon

3) How much time is there between when a star rises and when it sets?
   a. less than twelve hours
   b. about twelve hours
   c. more than twelve hours
   d. **it depends on the star**

4) From your position at 45° North latitude, you look straight north and see a star near the horizon. A little later you notice that it has shifted its position slightly. Which way did it move?
   a. **to the right (east)**
   b. to the left (west)
   c. up (rising)
   d. down (setting)

5) You go out tonight and see the brightest star in the constellation Orion rise at about 10 P.M. One week from now this star will rise at about
   a. **9:30 P.M.**
   b. 10:00 P.M. (i.e., any change will not be noticeable)
   c. 10:30 P.M.
   d. 10 A.M.

6) One night, you see the star Sirius rise at exactly 7:36 P.M. The following night it will rise at
   a. 7:36 P.M.
   b. 7:40 P.M.
   c. **7:32 P.M.**
   d. 8:36 P.M.

7) One evening at midnight, you observe Leo high in the southern sky. Virgo is to the east of Leo and Cancer is to the west. One month earlier, which of these constellations was high in the southern sky at midnight?
   a. Leo
   b. Virgo
   c. **Cancer**

8) How did the Pythagorean model of the stellar sphere account for the different paths the Sun takes across the sky in summer and winter?
   a. The rotation rate of the stellar sphere changed seasonally.
   b. The direction of rotation for the stellar sphere reversed from summer to winter.
   c. The direction of the stellar sphere's axis of rotation changed throughout the year.
   d. **The Sun's position on the stellar sphere changed throughout the year.**

9) On the day that shadows were known to reach the bottom of a well in Syene, Eratosthenes measured the shadow angle in Alexandria to be 7° from which he concluded that Earth was about 25,000 miles around. If he had made a mistake in his measurement and found the angle to be 5°, the value he would have calculated would have been
   a. less than 25,000 miles.
   b. **greater than 25,000 miles.**

10) Aristarchus' method for measuring the relative distances to the Moon and Sun was theoretically sound, but his result seriously underestimated the Sun's distance. This is because
   a. the Sun is actually 20 times farther from us than the Moon.
   b. the Sun and the Moon are never visible at the same time.
   c. the Moon and Sun do not actually move in the same plane.
   d. this ratio changes too much during the lunar cycle.
   e. **the angle is so close to 90° that even small errors in measurement produce huge errors in the calculated ratio.**

11) You observe a star rising directly to the east. When this star reaches its highest position above the horizon, where will it be?
   a. high in the northern sky
   b. high in the eastern sky
   c. **high in the southern sky**
   d. high in the western sky
   e. directly overhead

12) Stars that never appear to set are called circumpolar. As you move from Earth's equator toward the North Pole, the number of circumpolar stars
   a. **increases.**
   b. decreases.
   c. stays the same.

13) In the celestial sphere model of the sky, the Sun's position is constantly changing; the path that it follows is called the ecliptic. About long does it take the Sun to complete one "trip" around the ecliptic? (i.e., if the Sun is aligned with a certain star, how long will it be until it is again aligned with that same star?)
   a. 23 hours 56 minutes
   b. 24 hours
   c. 27 days
   d. **365 days**

14) In the celestial sphere model of the sky, the Sun's position along the western horizon at sunset changes because
   a. the Earth is stationary.
   b. the stars rotate with the celestial sphere.
   c. the tilt of the Earth changes throughout the year.
   d. **the position of the Sun along the ecliptic is constantly changing.**
   e. the Earth's rotation axis is inclined at 23.5 degrees.

15) What is the approximate date when the Sun's path along the ecliptic crosses Earth's equator on its way north?
   a. January 3
   b. **March 21**
   c. June 21
   d. September 22
   e. December 21

16) How often is the Sun directly over Earth's equator?
   a. once a day
   b. once a month
   c. **once every six months**
   d. once a year

17) If the Sun's motion along the ecliptic were reversed, how would its daily motion appear?
   a. **It would continue to rise in the east and set in the west.**
   b. It would now rise in the west and set in the east.

THE PTOLEMAIC MODEL

18) A planet moving in retrograde motion will, over the course of one night, move
   a. **east to west.**
   b. west to east.
   c. not at all, as planets do not move with the stars.
   d. randomly, as planets move differently than the stars.

19) A planet moving in retrograde motion will, over the course of several weeks, move in what direction compared to the background stars?
   a. **east to west**
   b. west to east

20) During retrograde motion, planets rise
   a. **in the east each night.**
   b. in the west each night.
   c. at the same time every night.

21) Epicycles are the circular loops that _____ make(s) during an orbit about _____.
   a. distant stars—Earth
   b. Earth—the Sun
   c. **planets other than Earth—Earth**
   d. planets other than Earth—the Sun

22) Ptolemy's model contains the ad hoc assumption that the epicycles of Mercury and Venus are lined up with the Sun. This was to explain the observation that
   a. these two planets display retrograde motion.
   b. Venus and Mercury go through phases just like the Moon.
   c. **these planets are always observed near the Sun.**
   d. Mercury and Venus do not appear to move at a constant rate.
   e. these are the closest planets to the Earth.

23) Ptolemy's model was designed primarily to
   a. improve upon the Pythagorean model's explanation for the Sun's motion.
   b. explain why stellar parallax could not be observed.
   c. **account for the motions of the planets.**

24) Which of the following was <u>not</u> a feature of the Ptolemaic model?
   a. Planets ride on epicycles.
   b. **Earth's rotation governs the rate at which the deferents revolve.**
   c. Earth is not in the exact center of the deferents.
   d. Earth is surrounded by a stellar sphere.
   e. The epicycles of Mercury and Venus are always lined up with the Sun.

25) Ptolemy's model was able to account for the retrograde motion of planets. His model also predicted that the distances to the planets would change as they orbited Earth. When a planet was in retrograde motion, what did Ptolemy's model predict about the distance to the planet?
   a. The planet was at its maximum distance from Earth.
   b. The planet was at its average distance from Earth (i.e., neither a maximum nor a minimum).
   c. **The planet was at its minimum distance from Earth.**

26) In a strict sense, Ptolemy's model violated many of the fundamental features of the Pythagorean model. Which feature of the Pythagorean model did the Ptolemaic system manage to retain in a <u>strict sense</u>?
   a. **Stars reside on a stellar sphere.**
   b. Earth is at the exact center of the heavens.
   c. Heavenly bodies move along perfect circles centered on Earth.
   d. Heavenly bodies move at constant speed along their paths.

COPERNICUS THROUGH NEWTON

27) What aspect of the Ptolemaic model did Copernicus retain in his model?
    a.   The Earth is at the center of the heavens.
    b.   The Sun is at the center of the heavens.
    c.   **Planetary motion is governed by epicycles.**
    d.   We live in a geocentric universe.

28) The general heliocentric approach of Copernicus won favor among astronomers primarily because it
    a.   explained why we do not feel the Earth move.
    b.   placed the Earth at the center of the heavens.
    c.   **was more aesthetically pleasing than the old model.**
    d.   made more accurate predictions than the Ptolemaic model.

29) Copernicus was dissatisfied with the Ptolemaic model because
    a.   he could not visualize it.
    b.   he did not believe in the stellar sphere.
    c.   **it seemed overly contrived.**
    d.   it did not agree with observations.

30) Tycho Brahe proposed a geocentric model in which all the planets but Earth orbited the Sun. Could this have been ruled out by Galileo's observation of the phases of Venus?
    a.   yes
    b.   **no**

31) Kepler's second law (equal areas in equal times) says in effect that
    a.   planets move at a steady pace about the Sun.
    b.   the more remote planets (whose orbits sweep out larger areas) must orbit faster, to sweep out the whole area in a similar time.
    c.   planets move slower as they near the Sun.
    d.   **planets move faster as they near the Sun.**

32) Earth's speed changes as is orbits the Sun. Which of Kepler's laws describes this?
    a.   I (describes shape)
    b.   **II (law of equal areas)**
    c.   III (relates period to radius)

33) That Pluto takes longer to orbit the Sun than Earth does is a described by which of Kepler's laws?
   a. first law
   b. second law
   c. **third law**

34) The number of days between the vernal equinox and the autumnal equinox (our spring and summer) is somewhat greater than the time between the autumnal equinox and the vernal equinox (our fall and winter). This is a result of
   a. the precession of the equinoxes.
   b. **Kepler's second law (equal areas).**
   c. Kepler's third law (how periods relate to orbital radii).
   d. the use of leap years to correct our calendar.
   e. the tilt of our rotation axis compared to our axis of revolution.

35) If we lived in a solar system with only one planet, which of Kepler's laws could never be discovered observationally?
   a. first law
   b. second law
   c. **third law**

36) Consider two planets, A and B, orbiting a distant star. Planet B orbits twice as far from the star as Planet A does. How does Planet B's orbital period compare to Planet A's?
   a. half as long
   b. the same
   c. twice as long
   d. **more than twice as long**

37) Galileo's observation that was in immediate and direct contradiction to the Ptolemaic (epicycle) model of the solar system was
   a. **the phases of Venus.**
   b. spots on the Sun.
   c. craters on the Moon.
   d. that the Milky Way is composed of stars.

38) Venus is observed in what we could describe as first quarter phase (i.e., it appears half lit and half dark). Venus, therefore,
   a. is about as close to Earth as possible (near side of orbit).
   b. **appears about as far from the Sun as possible (i.e., it is separated from the Sun by a large angle).**
   c. is about as far from Earth as possible (far side of orbit).

39) Venus is observed in a very narrow crescent phase (i.e., like the Moon only days after it's new). Venus, therefore,
   a. **is about as close to Earth as possible (near side of orbit).**
   b. appears about as far from the Sun as possible (i.e., it is separated from the Sun by a large angle).
   c. is about as far from Earth as possible (far side of orbit).

40) Newton introduced the universal law of gravity, and with it was able to explain
   a. **all three of Kepler's laws.**
   b. why no one could detect stellar parallax.
   c. why Mars moved in "retrograde loops" as seen from Earth.

41) The time it takes a small body to orbit about a large one depends only on
   a. the mass of the small body.
   b. the mass of the large body.
   c. the radius of the orbit.
   d. a and b
   e. **b and c**

42) If our Moon were suddenly to split into two pieces with one piece twice as massive as the other, which of the following would be true of the orbits the two pieces would make about the Earth?
   a. The large one would orbit twice as fast as the small one.
   b. The small one would orbit twice as fast as the large one.
   c. Both orbits would remain unchanged.
   d. The large piece would move closer to Earth and the small piece would move farther away.
   e. The small piece would move closer to Earth and the large piece would move farther away.

43) If a small weather satellite is orbiting Earth at an altitude very far above the surface and the large International Space Station is orbiting Earth at an altitude quite close to the surface, what can you say about the length of time it takes each object to orbit Earth once if neither is using rocket propulsion?
   a. The nearby, large space station has a longer orbital period.
   b. **The distant, small weather satellite has a longer orbital period.**
   c. They have the same orbital period.
   d. Neither can orbit if they are not using rocket propulsion.

44) If a small weather satellite and the large International Space Station are orbiting Earth at the same altitude above Earth's surface, which of the following is true?
   a. The large space station has a longer orbital period.
   b. The small weather satellite has a longer orbital period.
   c. **They have the same orbital period.**

STELLAR PARALLAX

45) Star X is known to be 10 parsecs away from us and star Y is 50 parsecs away. Which star has the greater parallax angle?
   a. **Star X**
   b. Star Y
   c. Neither—their parallax angles are the same.

46) A nearby star with parallax appears to move back and forth compared with background stars over the course of a year. This looks different from the motion of stars around the center of the Galaxy ("proper motion") because
   a. Stars with parallax never have proper motion.
   b. Stars with proper motion show redshift, while those with parallax do not.
   c. **Stars without parallax do not appear to reverse direction.**
   d. Proper motion is not detectable.

47) You observe two stars over the course of a year (or more) and find that both stars have measurable parallax angles. Star X has a parallax angle of 1 arcsecond. Star Y has a parallax angle of 0.5 arcsecond. Which star is closer?
   a. **Star X**
   b. Star Y
   c. There is insufficient information to determine this.

48) Which of the following stars is closest to us?
   a. Procyon (parallax angle = 0.29 arcsecond)
   b. Ross 780 (parallax angle = 0.21 arcsecond)
   c. Regulus (parallax angle = 0.04 arcsecond)
   d. **Sirius (parallax angle = 0.38 arcsecond)**

49) On Earth, the parallax angle measured for the star Procyon is 0.29 arcseconds. If you were able to measure Procyon's parallax angle from Venus, what would it be? (<u>Note</u>: Earth's orbital radius is larger than Venus's orbital radius.)
   a. more than 0.29 arcsecond
   b. 0.29 arcsecond
   c. **less than 0.29 arcsecond, but not zero**
   d. zero arcseconds (no parallax)

50) Consider two stars (X and Y). If star X is 3 parsecs away and star Y is 5 parsecs away, which has the greater parallax angle?
   a. **Star X**
   b. Star Y
   c. not enough information

51) Imagine that life did indeed develop on Mars and, at least for a time, there were beings there who were actively interested in astronomy. Further, they also used the method of stellar parallax to measure the distance to nearby stars using exactly the same definition of the parsec that we use—call it the marparsec (Martian parsec). Compared to our parsec, the marparsec would be
   a. **longer.**
   b. shorter.
   c. the same.
   d. We have no way of knowing this.

APPARENT AND ABSOLUTE MAGNITUDES

52) Rigel has apparent magnitude +0.1 and Acrux has apparent magnitude +1.4. Which star appears brighter in the sky?
   a. **Rigel**
   b. Acrux
   c. There is insufficient information to determine this.

53) Star G has an apparent magnitude of +5.0 and an absolute magnitude of +4.0. Star H has an apparent magnitude of +4.0 and an absolute magnitude of +4.0. Which star will appear brighter in the night sky?
   a. Star G
   b. **Star H**
   c. They will appear the same.

54) Star G has an apparent magnitude of +5.0 and an absolute magnitude of +4.0. Star H has an apparent magnitude of +4.0 and an absolute magnitude of +4.0. Which star is emitting the greater amount of light energy?
   a. Star G
   b. Star H
   c. **They are the same.**

55) Imagine that you are viewing a star that has an apparent magnitude of +0.2 and is located about 60 parsecs away from us. Which of the following is the most likely absolute magnitude for this star?
   a. **−3.7**
   b. +0.1
   c. +0.2
   d. +0.3
   e. +3.7

56) Capella is at a distance of 14 parsecs and has apparent magnitude +0.05. Which of the following could reasonably be Capella's absolute magnitude?
   a. +4.05
   b. −10.5
   c. **−0.7**
   d. +0.85

57) Star A has an apparent magnitude of +1.0 and an absolute magnitude of +1.0. If the distance between Earth and the star were increased, the apparent magnitude number would _____ and the absolute magnitude number would _____.
   a. increase—decrease
   b. decrease—increase
   c. **increase—not change**
   d. not change—increase
   e. decrease—not change

58) Altair's apparent magnitude is +0.77 and its absolute magnitude is +2.2. What is Altair's distance from Earth?
   a. **less than 10 parsecs**
   b. 10 parsecs
   c. more than 10 parsecs

59) Star F is known to have an apparent magnitude of –26 and an absolute magnitude of +5. The distance to Star F is
    a.   much more than 10 parsecs.
    b.   slightly more than 10 parsecs.
    c.   about 10 parsecs.
    d.   slightly less than 10 parsecs.
    e.   **much less than 10 parsecs.**

60) Spica is located about 80 parsecs away. Spica's absolute magnitude is –3.6. Which of the following is most likely Spica's apparent magnitude?
    a.   –8.1
    b.   –4.2
    c.   –3.6
    d.   **+0.9**
    e.   +3.6

LIGHT AND SPECTRA

61) The following graph shows the blackbody spectra for three different stars. Which of the stars is at the highest temperature?
    a.   **Star A**
    b.   Star B
    c.   Star C

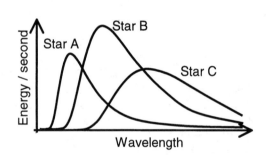

62) Imagine that you observe the Sun while in your space ship far above Earth's atmosphere. Which of the following spectra would you observe by analyzing the sunlight?
    a.   **dark line absorption spectrum**
    b.   bright line emission spectrum
    c.   continuous spectrum

63) Imagine that you are on the surface of Earth (below the atmosphere) and are observing the Sun. Which of the following spectra would you observe by analyzing the sunlight?
    a.   **dark line absorption spectrum**
    b.   bright line emission spectrum
    c.   continuous spectrum

64) Consider the dark line absorption spectra shown below for Star X and Star Z. What can you determine about the relative temperatures of the two stars?

Star X   Star Z

   a. Star X is at the higher temperature.
   b. Star Z is at the higher temperature.
   c. Both stars are the same temperature.
   d. **The relative temperatures of the stars cannot be determined.**

65) Consider the spectral curves for Star V and Star Y shown in the following graph. What can you determine about the relative temperatures of the two stars?
   a. **Star V is at the higher temperature.**
   b. Star Y is at the higher temperature.
   c. Both stars are the same temperature.
   d. The relative temperatures of the stars cannot be determined.

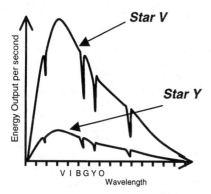

66) The stars Antares and Mimosa each have absolute magnitude −4.6. Antares is spectral type M and Mimosa is spectral type B. Which star is larger?
   a. **Antares**
   b. Mimosa
   c. They are the same size.
   d. There is insufficient information to determine this.

67) Rigel is about 100,000 times as luminous as the Sun and belongs to spectral class B8. Sirius B is about $^1/_{3000}$ times as luminous as the Sun and belongs to spectral class B8. Which star has the greater surface temperature?
   a. Rigel
   b. Sirius B
   c. **They have the same temperature.**
   d. There is insufficient information to determine this.

68) How does the size of a star near the top left of the H-R diagram compare with a star of the same luminosity near the top right of the H-R diagram?
   a. They are the same size.
   b. The star near the top left is larger.
   c. **The star near the top right is larger.**
   d. There is insufficient information to determine this.

69) How does the size of a star near the top left of the H-R diagram compare with a star at the same temperature near the bottom left of the H-R diagram?
   a. They are the same size.
   b. **The star near the top left is larger.**
   c. The star near the bottom left is larger.
   d. There is insufficient information to determine this.

70) You observe two stars with the same absolute magnitude and determine that one is a spectral type A star while the other is a spectral type F star. Which star has the greater surface area?
   a. the A star
   b. **the F star**
   c. The surface areas are the same.

71) Using spectroscopic parallax, you find a star's distance to be 76 parsecs. You now find out that the star isn't a main sequence star—it's actually a red giant. This makes your distance estimate
   a. too large.
   b. **too small.**
   c. fine—no significant change in estimate is needed.

72) You forget that the star Betelgeuse is a red giant and apply the method of spectroscopic parallax to determine its distance. Is the star farther away or closer than you thought?
   a. closer
   b. **farther away**

73) You use the method of spectroscopic parallax to determine that star SBGC1997 is at a distance of 146 parsecs assuming that it lies in the center of the main sequence band. It turns out that SBGC1997 actually lies along the lower edge of the main sequence band. Its actual distance is therefore
   a. **less than 146 parsecs.**
   b. greater than 146 parsecs.

74) Recall that the spectroscopic sequence is O, B, A, F, G, K, M from hottest to coldest. You use the method of spectroscopic parallax to determine the distance to an F2 star as 43 parsecs. You later discover that the star has been misclassified and is actually a type G7. The actual distance to the star must therefore be
   a. **less than 43 parsecs.**
   b. greater than 43 parsecs.

75) Assume that you were to use the method of "standard candles" to measure the distance to a star. If you failed to account for the effects of interstellar dust, your estimate of the star's distance would tend to be
   a. too small.
   b. **too great.**
   c. unaffected by this oversight.

Stellar Formation and Evolution

76) How does the Sun produce the energy that heats our planet?
   a. The gases inside the Sun are on fire; they are burning like a giant bonfire.
   b. **Hydrogen atoms are combined into helium atoms inside the Sun's core. Small amounts of mass are converted into huge amounts of energy in this process.**
   c. When you compress the gas in the Sun, it heats up. This heat radiates outward through the star.
   d. Magnetic energy gets trapped in sunspots and active regions. When this energy is released, it explodes off the Sun as flares that give off tremendous amounts of energy.
   e. The core of the Sun has radioactive materials that give off energy as they decay into other elements.

77) Consider the following information about the lifetime of three main sequence stars A, B, and C.
  ▪ Star A will be a main sequence star for 45 billion years.
  ▪ Star B will be a main sequence star for 70 million years.
  ▪ Star C will be a main sequence star for 800 thousand years.
  Which star has the greatest mass?
  a. Star A
  b. Star B
  c. **Star C**
  d. Stars A, B, and C all have approximately the same mass.

78) Stars A and B are formed at about the same time. Star B has 3 times the mass of star A. Star A has an expected lifetime of 3 billion years. What is the expected lifetime of star B?
  a. more than 9 billion years
  b. about 9 billion years
  c. 3 billion years
  d. about 1 billion years
  e. **less than 1 billion years**

79) The most important property in determining the eventual fate of a star is
  a. **its mass.**
  b. the details of its chemical composition.
  c. its proximity to high mass stars near the center of the galaxy.
  d. the initial concentration of iron in its core.

80) The eventual fate of a low-mass star like our Sun is to become a
  a. neutron star.
  b. black hole.
  c. a or b.
  d. **white dwarf.**

81) A supernova is
  a. a nova event that occurs within 75 parsecs of our Sun and is therefore particularly bright and impressive.
  b. the eventual fate of a high-mass star.
  c. the result of a white dwarf adding enough material from a binary companion to exceed its Chandrasekhar limit.
  d. a or b.
  e. **b or c.**

82) For a white dwarf to become a nova it is necessary for it to
   a. **have a companion star (be a member of a binary).**
   b. exceed its Chandrasekhar limit.
   c. have begun life as a high-mass star.
   d. continue the fusion cycle until its core is completely composed of iron.

83) Which of the following events will <u>not</u> leave any remnant?
   a. **type I supernova**
   b. type II supernova (death of a high mass star)
   c. nova

84) The Sun will likely never become a nova because this only happens to stars
   a. much more massive than the Sun.
   b. much less massive than the Sun.
   c. **in close binary pairs.**
   d. that have no planetary systems.

85) Compared with the variations in luminosity between stars, the variations in mass are
   a. much larger.
   b. about the same.
   c. **much smaller.**

86) What powers a white dwarf?
   a. chemical combustion
   b. the conversion of matter to energy
   c. **heat that is left over from its formation**
   d. heating as the star settles under its own weight

87) You observe two similar star clusters, which you label X and Y. You collect enough spectroscopic data on X and Y to plot their H-R diagrams. You discover that the most luminous main-sequence star in X is nearly ten times brighter than in Y. What do you conclude from this?
   a. X is a larger cluster than Y.
   b. Y is a larger cluster than X.
   c. X is an older cluster than Y.
   d. **Y is an older cluster than X.**

88) If a white dwarf exceeds its Chandrasekhar limit, it will
    a. become a nova.
    b. **become a type I supernova.**
    c. eject a planetary nebula.
    d. collapse into a black hole.

89) A white dwarf will eventually become a black dwarf unless it
    a. **has a nearby binary companion.**
    b. is too hot.
    c. rejoins the main sequence.
    d. becomes a black hole.
    e. exhausts its supply of helium first.

SOLAR SYSTEM PROPERTIES AND EVOLUTION

90) Which pair of planets listed originally formed at a location that had a temperature above the boiling point of water?
    a. Mars and Jupiter
    b. Saturn and Jupiter
    c. Mercury and Saturn
    d. Venus and Jupiter
    e. **Earth and Mars**

91) Would it have been possible for a large, Jupiter-like planet to form in the location of Mars?
    a. yes
    b. **no**

92) The composition of which group of planets best reflects the composition of the nebula out of which the solar system formed?
    a. the terrestrial planets
    b. **the Jovian planets**

93) We believe that the denser planets tend to be located nearer the Sun because
    a. it is predicted by Newton's laws.
    b. the gravity of the Sun affects them more.
    c. **it was hotter near the Sun when these planets formed.**
    d. angular momentum is a conserved quantity.

94) Which of the following properties of the planets does <u>not</u> tell us something about the way in which our solar system must have evolved?
   a. The planets all orbit in about the same plane.
   b. **The planets farther from the Sun take longer to complete their orbits.**
   c. The Sun spins in the same sense as the planets orbit.
   d. The planets' orbits are all nearly circular.
   e. Most planets spin "upright."

GALACTIC DISTANCES AND HUBBLE'S LAW

95) Cepheid variable stars are located in two different galaxies, A and B. Both stars have that same average <u>apparent</u> magnitude. The star in galaxy A has a bright-dim-bright period of 10 days, while the one in galaxy B has a bright-dim-bright period of 30 days. Which of the two stars has the greater average luminosity?
   a. the one in galaxy A
   b. **the one in galaxy B**
   c. They are the same.
   d. There is insufficient information to determine this.

96) Which of the two galaxies described in the previous question is at a <u>greater</u> distance from us?
   a. galaxy A
   b. **galaxy B**
   c. They are located at the same distance.
   d. There is insufficient information to determine this.

97) One of the emission lines of hydrogen occurs at a wavelength of 6563 angstroms. You measure the absorption spectrum from a star and determine that this same line appears at a wavelength of 6440 angstroms. Is the star moving toward the Earth or away from it?
   a. **toward**
   b. away

98) Astronomers currently believe that the Hubble constant has a value of about 70 km/s/Mpc. If some new measurement revealed that instead the Hubble constant is closer to 200 km/s/Mpc, what would this imply about the age of the universe?
   a. **It is much younger than current estimates.**
   b. It is much older than current estimates.

99) You observe two different Cepheid variable stars, X and Y. Star X has a bright-dim-bright period of 5 days, while Star Y has a bright-dim-bright period of 18 days. Star Y appears brighter than Star X. Which star, if either, is located closer to Earth?
   a. Star X
   b. Star Y
   c. **The answer cannot be determined.**

100) You observe two different Cepheid variable stars, C and D. Star C has a bright-dim-bright period of 5 days, while Star D has a bright-dim-bright period of 18 days. Star C appears brighter than Star D. Which star, if either, is located closer to Earth?
   a. **Star C**
   b. Star D
   c. The answer cannot be determined.

101) The amount of time for a Cepheid variable to complete its light-dim-light cycle can used as a <u>direct</u> measure of
   a. temperature.
   b. distance.
   c. **luminosity.**
   d. age.

102) Consider three widely separated galaxies in an expanding universe. Imagine that you are located in galaxy 1 and observed that both galaxies 2 and 3 are moving away from you. If you asked observers in galaxy 3 to describe how galaxy 2 appears to move, what would they say?

1   2   3

   a. Galaxy 2 is not moving.
   b. Galaxy 2 is moving toward the observer in galaxy 3.
   c. **Galaxy 2 is moving away from the observer in galaxy 3.**

EARTH'S MOTION AND SEASONS

103) If we could suddenly double the speed at which Earth orbits the Sun while keeping its rotation rate constant, the length of the mean solar day would
   a. **get slightly longer.**
   b. not change.
   c. get slightly shorter.
   d. be cut in half.
   e. double.

104)   If we were to suddenly stop having leap years (i.e., all years were 365 days long), the date of the vernal equinox would continually change. After remaining in March for a while, the next month in which it would appear would be
   a. February.
   b. **April.**

105)   We need a leap year because
   a. Earth's axis is tilted.
   b. the direction in which Earth's axis points slowly changes with time.
   c. **Earth does not go once around the Sun in exactly 365 days.**
   d. the sidereal and solar days are not the same.

106)   When the Sun reaches the highest point on the ecliptic on about June 21, its light will be striking us as the largest possible angle above the horizon and the day will be the longest of the year. Why does the weather continue to get warmer after this date even though the days are shortening and the sunlight is getting less direct?
   a. We are still moving closer to the Sun.
   b. The Sun only "appears" to be at its highest point on June 21 but actually continues to move northward for at least another month.
   c. **Water, rock, and dirt take a long time to warm up or cool down.**
   d. Earth continues to tilt its top toward the Sun after the summer solstice.

107)   The summer solstice (about June 21) is the time when the Sun
   a. is closest to the Earth.
   b. **is at its most northerly position along the ecliptic.**
   c. crosses the plane of the equator on its way north along the ecliptic.
   d. crosses the plane of the equator on its way south along the ecliptic.
   e. has the greatest probability of being eclipsed by the Moon.

108)   Our summer is somewhat longer than our winter (defined by the times between the solstices not by the weather) because
   a. **the Earth travels somewhat faster when it is closer to the Sun.**
   b. we have a leap year once every four years.
   c. there is an odd number of days in the year.
   d. the Earth spins counterclockwise about its own axis.

MOON PHASES

109)   The following diagram shows Earth and the Sun as
well as five different possible positions for the Moon.
Which position of the Moon best corresponds with the
phase of the Moon shown in the figure at the right?
[ANS: D]

```
  Moon's orbit

         E
     A                        ☀
         Earth               Sun
     B
            D

         C
NOT TO SCALE
```

110)   When the Moon appears to completely cover the Sun (an eclipse),
the Moon must be at which phase?
   a. full
   b. **new**
   c. first quarter
   d. last quarter
   e. at no particular phase

111)   You observe a full moon rising in the east. Which image shown best
represents how the Moon will appear as it is setting?
[ANS: C]

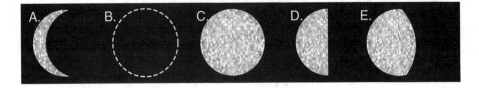

112)   It is just sunrise, and you are up early for a fishing trip. Your companion comments on the beautiful crescent moon. Which direction should you look to see the Moon?
a.   due south
b.   due north
c.   **southeast**
d.   southwest

113)   Which of the following is a possible observation at 4 A.M.?
a.   full Moon just rising in the east
b.   crescent Moon just setting in the west
c.   **crescent Moon rising in the east**
d.   gibbous Moon rising in the east
e.   first-quarter Moon in the south-west sky

114)   An astronaut is standing on the Moon when the Moon is in its waxing crescent phase. In what phase does Earth appear to the astronaut?
a.   full
b.   **gibbous**
c.   quarter
d.   crescent
e.   new

Appendix C: Astronomy Diagnostic Test[1]

1. As seen from your current location, when will an upright flagpole cast no shadow because the Sun is directly above the flagpole?
    A. Every day at noon
    B. Only on the first day of summer
    C. Only on the first day of winter
    D. On both the first days of spring and fall
    E. Never from your current location

2. When the Moon appears to completely cover the Sun (an eclipse), the Moon must be at which phase?
    A. Full
    B. New
    C. First quarter
    D. Last quarter
    E. At no particular phase

3. Imagine that you are building a scale model of the Earth and the Moon. You are going to use a 12-inch basketball to represent the Earth and a 3-inch tennis ball to represent the Moon. To maintain the proper distance scale, about how far from the surface of the basketball should the tennis ball be placed?
    A. 4 inches (1/3 foot)
    B. 6 inches (1/2 foot)
    C. 36 inches (3 feet)
    D. 30 feet
    E. 300 feet

4. You have two balls of equal size and smoothness, and you can ignore air resistance. One is heavy, the other much lighter. You hold one in each hand at the same height above the ground. You release them at the same time. What will happen?
    A. The heavier one will hit the ground first.
    B. They will hit the ground at the same time.
    C. The lighter one will hit the ground first.

5. How does the speed of radio waves compare to the speed of visible light?
    A. Radio waves are much slower.
    B. They both travel at the same speed.
    C. Radio waves are much faster.

---

[1] ©1999 The Collaboration for Astronomy Education Research (CAER). An electronic version is available at http://solar.physics.montana.edu/aae/adt/.

6. Astronauts inside the Space Shuttle float around as it orbits the Earth
    because
    A. there is no gravity in space.
    B. they are falling in the same way as the Space Shuttle.
    C. they are above the Earth's atmosphere.
    D. there is less gravity inside the Space Shuttle.
    E. more than one of the above.

7. Imagine that the Earth's orbit were changed to be a perfect circle about
    the Sun so that the distance to the Sun never changed. How would this
    affect the seasons?
    A. We would no longer experience a difference between the seasons.
    B. We would still experience seasons, but the difference would be
        much LESS noticeable.
    C. We would still experience seasons, but the difference would be
        much MORE noticeable.
    D. We would continue to experience seasons in the same way we do
        now.

8. Where does the Sun's energy come from?
    A. The combining of light elements into heavier elements
    B. The breaking apart of heavy elements into lighter ones
    C. The glow from molten rocks
    D. Heat left over from the Big Bang

9. On about September 22, the Sun sets directly to the west as shown on
    the following diagram. Where would the Sun appear to set two weeks
    later?
    A. Farther   B. In the   C. Farther
        south   same place   north

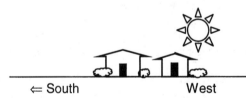

$\Leftarrow$ South   West   North $\Rightarrow$

10. If you could see stars during the day, this is what the sky would look like at noon on a given day. The Sun is near the stars of the constellation Gemini. Near which constellation would you expect the Sun to be located at sunset?
    A. Leo
    B. Cancer
    C. Gemini
    D. Taurus
    E. Pisces

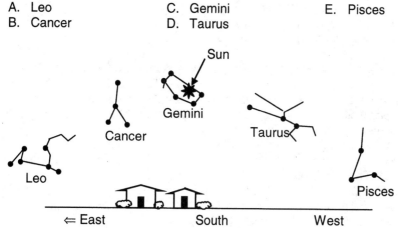

11. Compared to the distance to the Moon, how far away is the Space Shuttle (when in space) from the Earth?
    A. Very close to the Earth
    B. About half way to the Moon
    C. Very close to the Moon
    D. About twice as far as the Moon

12. As viewed from our location, the stars of the Big Dipper can be connected with imaginary lines to form the shape of a pot with a curved handle. To where would you have to travel to first observe a considerable change in the shape formed by these stars?
    A. Across the country
    B. A distant star
    C. Europe
    D. Moon
    E. Pluto

13. Which of the following lists is correctly arranged in order of closest-to-most-distant from the Earth?
    A. Stars, Moon, Sun, Pluto
    B. Sun, Moon, Pluto, stars
    C. Moon, Sun, Pluto, stars
    D. Moon, Sun, stars, Pluto
    E. Moon, Pluto, Sun, stars

14. Which of the following would make you weigh half as much as you do right now?
    A. Take away half of the Earth's atmosphere.
    B. Double the distance between the Sun and the Earth.
    C. Make the Earth spin half as fast.
    D. Take away half of the Earth's mass.
    E. More than one of the above

15. A person is reading a newspaper while standing 5 feet away from a table that has on it an unshaded 100-watt light bulb. Imagine that the table were moved to a distance of 10 feet. How many light bulbs in total would have to be placed on the table to light up the newspaper to the same amount of brightness as before?
    A. One bulb   D. Four bulbs
    B. Two bulbs   E. More than four bulbs
    C. Three bulbs

16. According to modern ideas and observations, what can be said about the location of the center of the Universe?
    A. The Earth is at the center.
    B. The Sun is at the center.
    C. The Milky Way Galaxy is at the center.
    D. An unknown, distant galaxy is at the center.
    E. The Universe does not have a center.

17. The hottest stars are what color?
    A. Blue
    B. Orange
    C. Red
    D. White
    E. Yellow

18. The following diagram shows the Earth and Sun as well as five different possible positions for the Moon. Which position of the Moon would cause it to appear like the picture at right when viewed from Earth?

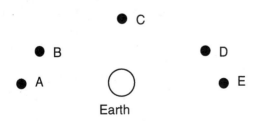

19. You observe a full Moon rising in the east. How will it appear in six hours?

A.   B.   C.   D.

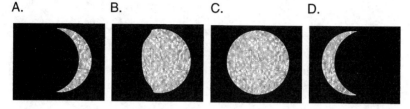

20. With your arm held straight, your thumb is just wide enough to cover up the Sun. If you were on Saturn, which is 10 times farther from the Sun than the Earth is, what object could you use to just cover up the Sun?
    A. Your wrist
    B. Your thumb
    C. A pencil
    D. A strand of spaghetti
    E. A hair

21. Global warming is thought to be caused by
    A. the destruction of the ozone layer.
    B. trapping of heat by nitrogen.
    C. addition of carbon dioxide.

22. In general, how confident are you that your answers to this survey are correct?
    A. Not at all confident (just guessing)
    B. Not very confident
    C. Not sure
    D. Confident
    E. Very confident

23. What is your college major (or current area of interest if undecided?)
    A. Business
    B. Education
    C. Humanities, Social Sciences, or the Arts
    D. Science, Engineering, or Architecture
    E. Other

24. What was the last math class you completed prior to taking this course?
    A. Algebra
    B. Trigonometry
    C. Geometry
    D. Pre-Calculus
    E. Calculus

25. What is your age?
    A. 0-20 years old
    B. 21-23 years old
    C. 24-30 years old
    D. 31 or older
    E. Decline to answer

26. Which best describes your home community (where you attended high school)?
    A. Rural   D. Urban
    B. Small town   E. Not in the USA
    C. Suburban

27. What is your gender?
    A. Female   C. Decline to answer
    B. Male

28. Which best describes your ethnic background?
    A. African-American
    B. Asian-American
    C. Native-American
    D. Hispanic-American
    E. None of the above (see question 29 below)

29. Which best describes your ethnic background?
    A. African (not American)
    B. Asian (not American)
    C. White, non-Hispanic
    D. Multicultural
    E. None of the above (see question 28 above)

30. How good at math are you?
    A. Very poor   C. Average   E. Very good
    B. Poor   D. Good

31. How good at science are you?
    A. Very poor   C. Average   E. Very good
    B. Poor   D. Good

32. Which best describes the level of difficulty you expect/experienced from this course?
    A. Extremely difficult for me   D. Easy for me
    B. Difficult for me   E. Very easy for me
    C. Unsure

33. How many astronomy courses at the college level have you taken?
    A. I'm re-taking this course.
    B. This is my first college-level astronomy course.
    C. This is my second college-level astronomy course.
    D. I've completed more than two other college-level astronomy courses.

Appendix D: Bloom's Taxonomy

TABLE 1: Bloom's Taxonomy of Educational Objectives for *Knowledge-Based Goals*

Level of Expertise	Description of Level	Example of Measurable Learning Objective
Knowledge	Recall, or recognition of terms, ideas, procedure, theories, etc.	When is the first day of spring?
Comprehension	Translate, interpret, extrapolate, but not see full implications or transfer to other situations, closer to literal translation.	What does the summer solstice represent?
Application	Apply abstractions, general principles, or methods to specific concrete situations.	Why are seasons reversed in the southern hemisphere?
Analysis	Separation of a complex idea into its constituent parts and an understanding of organization and relationship between the parts. Includes realizing the distinction between hypothesis and fact as well as between relevant and extraneous variables.	What would Earth's seasons be like if its orbit was perfectly circular?
Synthesis	Creative, mental construction of ideas and concepts from multiple sources to form complex ideas into a new, integrated, and meaningful pattern subject to given constraints.	Given a description of a planet's seasons, what would you propose its orbital and tilt characteristics to be?
Evaluation	To make judgment of ideas or methods using external evidence or self-selected criteria substantiated by observations or informed rationalizations.	What would be the important, and irrelevant, variables for predicting seasons on a newly discovered planet?

Table 2: Bloom's Taxonomy of Educational Objectives for *Skills-Based Goals*

Level of Expertise	Description of Level	Example of Measurable Learning Objective
Perception	Uses sensory cues to guide actions	Student realizes a "fuzzy" object might be interesting to explore further.
Set	Demonstrates a readiness to take action to perform the task or objective.	Student states that a telescope would be the most appropriate tool for investigating a "fuzzy" object in the sky.
Guided Response	Knows steps required to complete the task or objective.	Student can describe the steps involved in setting-up and aligning a telescope, and using it to find objects in the sky.
Mechanism	Performs task or objective in a somewhat confident, proficient, and habitual manner.	Student can eventually locate three given galaxies.
Complex Overt Response	Performs task or objective in a confident, proficient, and habitual manner.	Student can easily, and accurately, locate three given galaxies.
Adaptation	Performs task or objective as above, but can also modify actions to account for new or problematic situations.	Student can locate three different galaxies on a partially cloudy night.
Organization	Creates new tasks or objectives incorporating learned ones.	Student can successfully design, and host, a star party using a telescope. Make modifications to your telescope to allow for mounting of a heavy CCD camera.

Table 3: Bloom's Taxonomy of Educational Objectives for *Affective Goals*

Level of Expertise	Description of Level	Example of Measurable Learning Objective
Receiving	Demonstrates a willingness to participate in the activity	When I'm in class I am attentive to the instructor, take notes, etc. I do not read the newspaper instead.
Responding	Shows interest in the objects, phenomena, or activity by seeking it out or pursuing it for pleasure	Student chooses to allocate free-time to watching the Discovery Channel™ instead of ESPN™.
Valuing	Internalizes an appreciation for (values) the objectives, phenomena, or activity.	Student believes that it is important that the local high school support an astronomy club.
Organizing	Begins to compare different values, and resolves conflicts between them to form an internally consistent system of values.	During vacations or business travels, student consistently includes side trips to local planetaria or astronomy exhibits.
Characterizing by a Value or Value Complex	Adopts a long-term value system that is "pervasive, consistent, and predictable."	Joins, recruits for, and regularly attends functions of, a local amateur astronomy club.

Appendix E: Collaborative Learning Tasks

Effective collaborative learning group tasks are specifically designed to encourage students to interact with and learn from each other. As opposed to review and homework questions, collaborative learning group tasks require students to make rationalized choices about what variables to explore, to compare their results, and to come to consensus. In most cases, various student groups will arrive at different results yet still be correct. It is this "multiple correct solutions" aspect to the tasks that make them useful for collaborative groups because the students have a variety of approaches and results worthy of discussion. Initially, students might feel uncomfortable addressing questions where there is more than one correct answer; regular use of appropriate collaborative learning group tasks will build a sense of confidence and community among students and create a learning environment where inquiry and curiosity is valued. Most are designed to take 10 to 15 minutes in the lecture classroom environment and should be used to break up a long lecture. Students should be encouraged to use their text as a resource as well as each other.

A common approach for assigning credit is to have one paper turned in per group with all group members sharing the same grade. This can significantly cut down on the number of papers to be graded. An additional strategy is to grade the papers using the following scheme: 10 points for a well thought out and extended response; 7 points for answering the question, but having less than optimal justification; 3 points for weak effort; and 0 points for no effort. It is important to make the collaborative group tasks a significant part of the course grade in order for them to be taken seriously by the students.

1. **Chasing Solar Eclipses**—Consider a text figure showing solar eclipse paths over a world map. As a group, write a description of which eclipse your group would most like to observe together, where and when you would go to observe it, and fully explain why you selected the date and site you did.
   Collaborative learning group activities work best when students generate ideas and articulate their reasoning in low-risk environments. Students should brainstorm the strengths and weaknesses of various observation sites and dates. They may need to be reminded of the import role played by weather. An emphasis on clearly explaining related costs and the vacation aspects of exotic travel for a 10-minute eclipse will encourage students to be creative in their justification. You never know; it is possible that some of your students will actually take an eclipse trip someday.

2. **Measuring Diameters from the Surface**—Eratosthenes used simple geometric reasoning to calculate Earth's size using shadows. As a group, create a sketch and an accompanying written description showing exactly how his measurements would lead to a different result using one of Jupiter's moons selected by your group.
    Students should use the physical data for Jupiter's moons from the text's appendix. This gives them an opportunity to explore the resources provided in the text as well as sketch and apply Eratosthenes's ideas in a novel environment. In every case for an object smaller than Earth, the differences in shadow lengths should be longer for similar arc lengths along the surface.

3. **Parallax Measurements**—If the angular width of your thumb at arm's length is about 1/2 of a degree, determine the angular size of four different objects in the room selected by your group members. Provide a sketch with an organized data table.
    A hallmark of appropriate collaborative group tasks is that different groups will end up with different, but still correct, answers. Encourage students to compare their results with other groups to emphasize this. If time allows, students can also complete this activity outdoors.

4. **Tourist Attraction on Sacred Ground**—Your group has been asked to arbitrate a dispute between a tour bus company and a nearby Native American tribe. The dispute centers on an ancient medicine wheel recently discovered by a team of university archeologists. Using sketches as necessary, compose a legal brief that describes what a medicine wheel is designed to do astronomically and summarize the opposing positions of the two groups.
    All too often, students think science operates independent of society. By asking students to focus on the intersection of science and society with authentic problems, students can come to understand that science is a human endeavor. This collaborative group activity might take longer than many others. Examples of these conflicts include construction of observatories near wildlife refuges and on mountain tops with significant cultural important (e.g., Mauna Kea in Hawaii).

5. **Galileo's Observations**—Your group should select what it believes to be Galileo's single most important astronomical observation. Explain what he observed using sketches and why it was most important.
    Galileo's observations had enormous societal and religious implications. By asking students to evaluate and sketch his observations, students can begin to appreciate his challenge in explaining the wonders he was

observing, including lunar craters, sunspots, countless unseen stars, and the Galilean moons.

6. **Radio Station Wavelength**—Determine the wavelength from your group's favorite radio station and calculate how long it takes for the radio waves to arrive at your current location from the radio station. *Calculations are a normal part of many astronomy courses that can become more meaningful if they involve aspects students are familiar with—such as their favorite radio station. Put closure on this activity by emphasizing the differences in groups' answers and the enormity of the electromagnetic spectrum.*

7. **Doppler Shift**—Calculate the new wavelength for a 2 kHz automobile horn would be for a car moving first toward you and then away from you at a speed agreed on by your group. *Calculations are a normal part of many astronomy courses that can become more meaningful if they involve aspects students are familiar with—such as the motion of vehicles. Emphasize the differences in students' answers as related to the various speeds the select.*

8. **Radio Interferometer**—Determine the maximum size interferometer your group could build if you placed 2-meter radio telescopes where each group member lives. *Students should estimate the distance between the homes of the two most distant group members, probably in miles, and then convert to appropriate units. Activities such as this encourage students to build a community of learners by learning more about each other.*

9. **Planetary Density**—Calculate the average density of any planet of your group's choosing, besides Earth, using the planetary radius and planetary mass as provided in the appendix of your textbook and compare it with Earth's average density. *Students should use the physical planetary data provided in the appendix to calculate and compare planet densities. Emphasize to students the importance of everyone in the group knowing how to do the calculation and to compare results with nearby groups.*

10. **Earth's Interior Scale Model**—Using a ruler and self-stick or taped-on labels, create a scale model of the Earth on the shortest member of your group. Use the group member's height in inches divided by Earth's diameter (12,800 km) as the scale factor along with figure XX *(find appropriate figure in text)*. For example, if the selected group member is 65 inches tall; then the 50-km maximum depth of Earth's crust is (65 inches / 12,800 km) x (50 km) = 0.25 inches from the top of the head and 0.25 inches from the bottom of the feet.

Most students have weak senses of size and scale and this activity focuses on helping students create appropriately sized scale models using a "scale factor." The concept of scale factors can be difficult for many students to grasp and should be approached slowly. CAUTION: You should be aware that this activity could make some students (and professors) uncomfortable and you should be extremely careful to avoid any unintentional references to individuals' body shapes and/or sizes.

11. **New Views of Mercury**—The craters on the Moon are named after great scientists and philosophers. As a group, propose new names for the 10 largest craters found on Mercury when its "other" side is imaged by the Mercury MESSENGER mission in 2007, and explain your reasoning.
   The naming of astronomical objects and their features is accomplished as a serious and time-consuming international effort of the International Astronomical Union. Students should be reminded that IAU names for objects and features named for individuals must be done after their deaths or 100 years later if they are political figures.

12. **Outpost on the Moon**—The anticipated cost of transporting a gallon of water from Earth to the Moon is $15,000. Estimate the cost of taking a single-day's supply of water for your group to the Moon by determining how much water each of the group members uses in a single day.
   Collaborative group learning provides opportunities for students to learn from one another and reflect on their own lives. Students will likely be quite surprised about how much clean water they use every day. Be sure they include flushing a toilet (up to 5 gallons), taking a shower (up to 25 gallons), cooking, cleaning, car washing, laundry, and so on.

13. **Space Travel**—Traveling at the spacecraft's speed of your group's choosing, calculate the minimum length of time it would take to travel from Earth to Venus.
   Students will struggle, possibly surprisingly, because the distance between Venus and Mars is not listed anywhere in their textbook. Encourage them to consider superior conjunction, opposition, or various positions in between to make their calculations easier. If you give them the positions or values up front, this activity becomes a simple calculation and is not worthy of group discussion.

14. **Planetary Years**—Use a planetary data appendix to determine how old, in Martian years, each member of your group would be if they were born on Mars.
   Even when faculty tell students that a year is the length of time for the Earth to orbit the Sun, they typically do not internalize this until they have to ask and answer questions related to this concept. This task is,

perhaps somewhat surprising, not a trivial task for many nonscience students.

15. **Evidence for Life**—As a group, decide on exactly what evidence you need in order to believe that life, as your group defines it, exists or existed on Mars.
    The traditional definition is that life is evidenced by ability to reproduce, adapt to an environment, alter an environment, and use energy. Some students ignore that the vast amount of life on Earth is in microbes and initially feel the need to see animals that walk around as evidence of life.

16. **Galilean Moons**—As a group, agree to which of the Galilean Moons you would send a robotic lander/rover mission and justify your choice.
    Each of the Galilean Moons have different characteristics worthy of study. Io has volcanoes, Europa has subsurface oceans, Ganymede has complex surface features, and Callisto holds clues to the formation of the Jovian moon systems. Ask students to fully justify what features are interesting and why they are worthy of study.

17. **Saturn's Rings Scale Model**—Using your tallest group member's outstretched arms, create a complete scale model of all Saturn's rings as listed on Table XX *(find appropriate figure in text)* using self-stick notes or tape labels. Measure the maximum distance from nose to fingertip and use this as the scale factor for the rings' maximum radius. For example, the outer radius of the E ring is 480,000 km and the inner radius of the D ring is 67,000 km; if the distance from nose to fingertip were 40 cm, the inner radius of the D ring would be 67,000 km × (40 cm / 480,000 km) = 5.6 cm from the nose.
    Most students have weak senses of size and scale and this activity focuses on helping students create appropriately sized scale models using a "scale factor." The concept of scale factors can be difficult for many students to grasp and should be approached slowly. CAUTION: You should be aware that this activity could make some students (and professors) uncomfortable and you should be extremely careful to avoid any unintentional references to individuals' body shapes and/or sizes.

18. **Scaled Distance to Pluto**—If the distance in meters between your group and the room's exit represents the average distance between the Sun and Earth, determine how far away the planets Uranus, Neptune, and Pluto would be located from where your group sits and ask three group members to stand at the appropriate distances.
    The concept of scale factors can be difficult for many students to grasp and should be approached repeatedly. Because students sit at varying distances to the room's exit, this activity results in "multiple correct

answers" which encourages students to evaluate their results more critically when comparing to other groups.

19. **Dinosaur Extinction**—Decide on what evidence would finally settle the "'extinction of the dinosaurs" debate once and for all for your group. *Students often think the scientific questions are decided by "majority rules." Students should come to consensus and provide a written description of what findings would settle the 'extinction of the dinosaurs' debate as described in the text.*

20. **Planet or Kuiper Belt Object?**—As a group, agree on whether or not Pluto should be classified a planet. Justify your answer. *The scientific community is undecided on whether Pluto should be classified as a planet. For historical reasons, it will likely continue to be known as a planet, but categorically, it is probably one of the larger members of the outer lying Kuiper Belt objects. Encourage students to justify an informed and defendable decision.*

21. **Solar Cycle**—Using figure XX *(find appropriate figure in text)* showing sunspot numbers over the last few hundred years, each member of your group should individually determine the number of years between any two sequential sunspot maxima. Average your group's results and compare the average value to the frequently stated 11-year solar cycle. *Students often develop the misconception that the solar cycle is exactly 11 years. Many students will think that they have erred in their data if they do not arrive at 11 years exactly, and this is a great opportunity to introduce the concept of variability.*

22. **Inverse Square Law**—Considering where your group is sitting right now, how many times dimmer would an imaginary, super-deluxe, ultra-bright flashlight be if it were located at the front door of the group member who lives farthest away as compared to if it were at the front door of the group member who lives closest? Explain your reasoning. *Many students struggle with the one over "r-squared" law, especially if they do not have an exact luminosity or brightness value provided. You can show students that the result will be the same regardless of the brightness value they choose.*

23. **Differences in Brightness**—As a group, select any two of stars in a text appendix listing of the sky's brightest stars and compare the apparent visual magnitudes to determine how many times brighter one is as compared to the other. *Using a difference of magnitude of one as approximately 2.5 times the energy and a difference of five as 100 times the energy, group members should come to agreement on which two stars to evaluate*

from the apparent visual magnitude of stars listed in most tables listing the sky's brightest stars to compare how many times brighter one is as compared to the other. Encourage students to approximate using integers.

24. **Exploring Density**—The interstellar medium has a very low density of about 1000 per cubic kilometer. Estimate the population density of students in the tallest dormitory on campus using units of students per cubic feet and compare to the population density of the classroom. Explain your reasoning.
Students traditionally do not conceptualize density. However, using characteristics of phenomena that they are more familiar with will help. To complete this back-of-the-envelop calculation, someone in the room should know how many rooms per floor and how many stories tall one of the campus dormitories is.

25. **Evolutionary Sequences**—As a plot of luminosity versus temperature, the H-R diagram is useful for describing how stars evolve over time even though "time" is not the label on either axis. As a group, create an imaginary graph of "dollars of financial income" (vertical axis) versus "weight" (horizontal axis) and use it to describe the past and future life cycle of one of your group members. Clearly label your diagram and provide a figure caption clearly explaining each life-phase.
The H-R diagram is an odd representation for many students to deal with. This task is designed to help "find" the time dimension by having them construct a similar graph for a familiar situation. Students need to label the diagram clearly and provide figure captions clearly explaining each life phase. Phases students might be encouraged to use include a childhood with almost no income, college years with negative income, working years with salary income, and retirement years with investment income. Creativity counts!

26. **Supernova Brightness**—Each group member should select a different star listed in an appendix table of the sky's brightest stars and determine what its new apparent visual magnitude would be IF it were to become a supernova and increase its brightness 10,000 times.
Remembering that a change in 5 magnitudes represents an energy change of 100 times, 100 x 100 is a change in 10 magnitudes. A star with an initial magnitude of 6 becomes a –4 in magnitude.

27. **Neutron Star Sizes**—The typical neutron star is only about 20 km in diameter. How many neutron stars could fit between the birthplaces of two members in your group? Explain your reasoning.
Students do not often internalize values provided in metric. You can help students understand size and scale of astronomical objects by

*helping them relate them to their world experiences. To determine
approximately how many 20-km neutron stars will fit in the distance
between birthplaces, students will likely have to roughly determine the
distances in miles and convert either the distance or the 20 km to U.S.
units. Remind students that they can use very rough numbers for this
back-of-the-envelope calculation.*

28. **Cepheid Pulsation Periods**—For each member of your group,
    determine the luminosity of stars that have pulsation periods equal to
    the number of days since each person last ate rice, using a text figure
    showing a graph of brightness versus time.
    *Many students do not closely relate graphical information into real life-
    time scales. Although any life event would be appropriate to use to
    consider with a graph, the high variability of a food item such as rice will
    provide students with a wide range of Cepheid luminosities.*

29. **Classification Schemes**—Edwin Hubble classified galaxies based on
    their appearance into three broad categories. Classify all the books
    used this term by the group member who is taking the most classes
    using categories such as color, size, thickness, and cost, and using any
    scheme that adequately describes the collection. Clearly define each
    category.
    *This activity is not about finding the correct answer but rather about the
    process of classification. Emphasize to students that clearly defined
    categories are the goal of any classification scheme rather than "which
    scheme is correct." You can follow up by discussing why some
    schemes are sustained while others fall into disuse.*

30. **Hubble's Constant**—For each member in your group, determine the
    maximum age of the universe if Hubble's constant value is three times
    each person's age.
    *Students do not often internalize values or relationships presented in
    class unless they have to use these in various contexts, such as ones
    related to their ages.*

31. **Definition of Life**—As a group, compose a paragraph everyone agrees
    with that defines life. It should clearly show that rocks are not alive and
    that plants are alive. According to your definition, are stars alive?
    Compare and contrast your group's definition with that from another
    group.
    *Life is often described as organisms that can react to their environment
    and often heal themselves when damaged; can grow by taking in
    nourishment from their surroundings and process it into energy;
    reproduce and pass along some of their own characteristics to their
    offspring; and have the capacity for genetic change and can therefore*

evolve from generation to generation so as to adapt to a changing environment. Even such a carefully constructed definition can unclearly categorize stars, fire, certain crystals, and, in particular, viruses, which are often not classified as being alive.

32. **The Drake Equation**—Independently, each person in your group should estimate the average lifetime of a technologically competent civilization as described in the Drake equation. Explain the variation in the values your group had.
    One of the more interesting variables in the Drake equation is the one related to the length of time before a technologically advanced civilization destroys itself. This is also the one over which society has the most control. Students often forget that the Roman Empire lasted only 500 years and that the United States is only 200 years old.

33. **Extraterrestrial Intelligence**—If your group were appointed to "speak for Earth" upon communication with an extraterrestrial, what would your group say? Write your group's speech and annotate it with explanations of why you choose to say what you do.
    If students seriously consider what they want "those who speak for us" to say to aliens, they might be uncomfortable with what our elected officials might say. This activity will provide students an opportunity to think about how life on Earth and astronomy do impact their own lives.

Appendix F: Sample Learning Objectives

1. Students will compare the characteristics amongst different types of telescopes.
   - know the three functions of a telescope and their relative importance
   - compare and contrast a reflector and refractor
   - know the important features to consider in buying a telescope
2. Students will know and understand terms that describe the night sky and its motions.
   - important terms: constellation, asterism, zodiac, circumpolar star, seasonal star
   - know the naming conventions for stars
   - understand the nightly and seasonal motions of stars
   - identify important seasonal constellations and asterisms on a star map
   - locate Polaris on a star map
3. Students will understand stellar magnitudes.
   - know how the apparent magnitude scale is defined
   - know how distance and luminosity affect apparent magnitude
   - know how absolute magnitude is defined
   - understand the relationships between apparent magnitude, absolute magnitude and distance
   - understand and apply the method of standard candles to estimate distances
4. Students will apply the procedure of trigonometric parallax to measure distance.
   - know how apparent motion depends on distance
   - understand the difficulties in using this process for measuring stellar distances
   - understand how shifts in apparent position are related to distance
   - know the definition of a parsec
5. Students will understand how spectra are used to classify stars and to determine their compositions and temperatures.
   - understand the concept of electromagnetic (EM) spectra
   - understand the relationship between spectral films and brightness versus wavelength graphs
   - recognize continuous, emission, and absorption spectra
   - understand the physical processes leading to the three types of spectra
   - know how blackbody spectra are used to find temperature
6. Students will use the H-R diagram to characterize stars.
   - understand that stars are classified according to a spectral scheme

- know the relationship between spectral class and temperature
- understand the relationship between luminosity, temperature, and size
- know the parts of an H-R diagram
- understand the scales and axes used on the H-R diagram
- know the names and properties of the major types of stars on the H-R diagram
- understand how the properties of star types are determined
- apply all of the above to describe a star

7. Students will apply the process of *spectroscopic parallax* for determining relative stellar distances.
   - know the steps in the procedure
   - identify the underlying assumptions and rationale for each step in the procedure
   - understand the importance of the H-R diagram in extending the distance ladder

8. Students will comprehend the different processes by which stars are born, evolve, and eventually die.
   - know how raw materials combine to form stars
   - know the basic life cycles of both low and high mass stars
   - understand why mass is so important in determining a star's fate
   - understand what determines whether the death of a star will result in a black hole, a neutron star, or a white dwarf
   - know the state of observational evidence for black holes, neutron stars, and white dwarfs
   - understand how a white dwarf can continue to evolve into either a nova or type I supernova and what condition determines this

9. Students will understand the overall structure of our universe.
   - know the basic size and structure of our own galaxy and the Sun's place in it
   - apply Hubble's classification scheme to identify to which of the four basic types a galaxy belongs
   - understand how Cepheid variables were used to extend the distance ladder to galaxies
   - know that galaxies are organized into larger structures called clusters and superclusters
   - know the two principle galaxies in the local cluster

10. Students will understand the observational evidence supporting the expanding universe hypothesis.
    - know how recessional (radial) velocities are determined
    - understand the relationship between recessional velocity and distance
    - understand how Hubble's constant determines the age of the universe

- understand what makes Hubble's constant difficult to determine
- understand why Quasars are of interest to astronomers
- know the key observations supporting the Big Bang Theory

11. Students will understand how the Greeks were able to learn about the size and shape of the Earth as well as the relative distances to the moon and Sun.
    - provide evidence that the Earth is round
    - understand the essential features of Eratosthenes' method for determining the size of the Earth
    - provide evidence that the Sun is farther away than the Moon
    - understand how Aristarchus predicted the relative distances to the Moon and Sun
    - apply Aristarchus' method to make predictions of relative Moon-Sun distances

12. Students will understand how Greek cosmological models accounted for available observations.
    - know what observational evidence was available to Greek astronomers including the motions of the stars, Sun, and Moon
    - know the times and meanings of the equinoxes and solstices
    - know how the Sun appears to move relative to the background stars
    - understand the meaning of the ecliptic
    - know the basic assumptions of the Pythagorean model
    - understand how the Pythagorean model accounts for most observations

13. Students will understand how the Ptolemaic model developed as a means for explaining the motions of the planets.
    - identify the shortcomings of the Pathagorean model that led to more sophisticated models
    - know the meaning of epicycles and deferents and how they were used to explain retrograde motion
    - identify the elements of the full Ptolemaic model that were really violations of the Greek ideals espoused by the Pythagoreans
    - given a deferent and epicycle, trace Ptolemaic planetary orbits as seen from the North Star

14. Students will understand the major contributions of Copernicus, Tycho, and Kepler.
    - explain why Copernicus was dissatisfied with the Ptolemaic model
    - identify the key elements of the Copernican system
    - identify the elements of the Greek system that were retained in the Copernican system
    - understand the significance of Tycho's discovery of a supernova and a comet
    - describe Tycho's major contributions to astronomy

- understand how to use Kepler's method to determine Mars's orbit from data listing the bearings to Mars taken at one-Martian-year intervals
- understand the meaning of Kepler's three laws of planetary motion

15. Students will understand the implications of Galileo's major observations.
    - know Galileo's seven important observations
    - understand how these observations contradicted the Greek ideal of the "perfect heavens"
    - understand how these observations were used to support the heliocentric model

16. Students will understand the major contributions of Isaac Newton.
    - know that objects in orbit are falling towards the center of the Earth
    - understand the connection between the falling of an apple and the "falling" of the Moon
    - understand the relationship between Newton's law of gravity and Kepler's laws of planetary motion

17. Students will understand how theory and observations worked together to identify the remaining planets in the solar system.
    - know what the Titius-Bode rule is and how it applies to the solar system
    - know the details of Herschel's discovery of Uranus
    - understand how the observed motion of Uranus led to the discovery of Neptune
    - understand how the discrepancies in the observed motion of Mercury were originally accounted for
    - know what led astronomers to search for the planet Pluto
    - know how "blinking" was used in the discovery of Pluto

18. Students will be able to describe the positions of astronomical objects from both a heliocentric and a geocentric perspective.
    - know how the observer's position on Earth makes particular objects in the sky visible at specific times
    - analyze the rotation of an Earth observer to predict the rising and setting times of sky objects
    - interpret heliocentric positions in terms of a geocentric model and vice versa

19. Students will know the basic physical and dynamical properties of the planets.
    - know the common features of the motions of the planets and moons
    - distinguish between the terrestrial and Jovian planets
    - know how the current model of planetary formation accounts for these properties

20. Student will know the properties of meteors, asteroids, and comets.

- identify the cause of meteors
- know the cause of meteor showers
- understand the best time for observing a meteor shower
- know where most asteroids reside
- know what distinguishes an Apollo asteroid
- know how asteroids are believed to affect life on Earth
- know the basic structure of comets
- understand what produces a comet's tail
- know the sources of short and long period comets

21. Students will have some basic knowledge of our star: the Sun.
    - know the approximate size, temperature, and composition of the Sun
    - know what sunspots are and how the solar cycle operates
    - know how, besides being the source of light and heat, the Sun affects life on Earth

22. Students will develop a mental model of the Sun-Earth-Moon geometry that both explains and predicts lunar phases.
    - know the terminology associated with the Moon phases
    - predict the phase of the Moon given the relative positions of the Sun-Earth-Moon system
    - predict the relative positions of the Sun-Earth-Moon system given the Moon phase and time of day

Appendix G: Attitude Survey

The following pre-survey, which is available along with the corresponding post survey at

> http://www.wcer.wisc.edu/nise/cl1/flag/tools/attitude/astpr.htm

was designed by Mike Zeilik at the University of New Mexico. It was modeled on a survey of students' attitudes toward introductory statistics developed by C. Shau. The items fall into four subscales: affect (attitude), cognitive competence, value, and difficulty. Affect relates to positive and negative attitudes about astronomy and science (8 items). Cognitive competence describes attitudes about the students' intellectual knowledge and skills when applied to astronomy and science (9 items). Value involves attitudes about the usefulness, relevance, and worth of astronomy and science in personal and professional life (9 items). Difficulty entails attitudes about the difficulty of astronomy and science as subjects (8 items).

Instructions—The questions below are designed to identify your attitudes about astronomy and science. Please read each question. From the 5-point scale mark the response that most clearly represents your agreement with the statement. Use the entire 5-point scale. Try not to think too deeply about each response; there are no correct or incorrect answers.

	Strongly disagree		Neither agree nor disagree		Strongly agree
1. Astronomy is a subject learned quickly by most people.	1	2	3	4	5
2. I will have trouble understanding astronomy because of how I think.	1	2	3	4	5
3. Astronomy concepts are easy to understand.	1	2	3	4	5
4. Astronomy is irrelevant to my life.	1	2	3	4	5
5. I will get frustrated going over astronomy tests in class.	1	2	3	4	5
6. I will be under stress during astronomy class.	1	2	3	4	5
7. I will understand how to apply analytical reasoning to astronomy.	1	2	3	4	5

8. Learning astronomy requires a great deal of discipline.   1   2   3   4   5

9. I will have no idea of what's going on in astronomy.   1   2   3   4   5

10. I will like astronomy.   1   2   3   4   5

11. What I learn in astronomy will not be useful in my career.   1   2   3   4   5

12. Most people have to learn a new way of thinking to do astronomy.   1   2   3   4   5

13. Astronomy is highly technical.   1   2   3   4   5

14. I will feel insecure when I have to do astronomy homework.   1   2   3   4   5

15. I will find it difficult to understand astronomy concepts.   1   2   3   4   5

16. I will enjoy taking this astronomy course.   1   2   3   4   5

17. I will make a lot of errors applying concepts in astronomy.   1   2   3   4   5

18. Astronomy involves memorizing a massive collection of facts.   1   2   3   4   5

19. Astronomy is a complicated subject.   1   2   3   4   5

20. I can learn astronomy.   1   2   3   4   5

21. Astronomy is worthless.   1   2   3   4   5

22. I am scared of astronomy.   1   2   3   4   5

23. Scientific conclusions are rarely presented in everyday life.   1   2   3   4   5

24. Scientific concepts are easy to understand.   1   2   3   4   5

25. Science is not useful to the typical professional.   1   2   3   4   5

26. The thought of taking a science course scares me.   1   2   3   4   5

27. I like science.   1   2   3   4   5

28. I find it difficult to understand scientific concepts.   1   2   3   4   5

29. I can learn science.   1   2   3   4   5

30. Scientific skills will make me more employable.   1   2   3   4   5

31. Science is a complicated subject.   1   2   3   4   5

32. I use science in my everyday life.   1   2   3   4   5

33. Scientific thinking is not applicable to my life outside my job.   1   2   3   4   5

34. Science should be a required part of my professional training.   1   2   3   4   5

Appendix H: Astronomy Education Research Annotated Bibliography

J. P. Adams and T. F. Slater, "Using Action Research to Bring the Large Lecture Course Down to Size," *Journal of College Science Teaching*, 28(2), 87—90 (1998).
- A series of astronomy action research studies including a high correlation between student self-report knowledge and examination scores.

J. P. Adams and T. F. Slater, "Astronomy in the National Science Education Standards," *Journal of Geoscience Education*, 48(1), 39—45 (2000).
- Literature review of astronomy education research organized by the NSES content learning objectives.

R. K. Atwood and V. A. Atwood, "Preservice Elementary Teachers' Conceptions of the Causes of Seasons," *Journal of Research in Science Teaching*, 33(5), 553—563 (1996).
- The authors propose that misconceptions about seasons might not be as firmly entrenched as commonly thought.

J. Baxter, "Children's Understanding of Familiar Astronomical Events," *International Journal of Science Education*, 11, 502—513 (1989).
- Many student conceptions are consistent across cultures.

P. L. Callison and E. L. Wright, "The Effect of Teaching Strategies Using Models on Preservice Elementary Teachers' Conceptions about Earth-Sun-Moon Relationships." ERIC Document ED 360 171 (1993).
- Students who use physical models during instruction have larger achievement gains than students who only use mental models.

A. Caramaza, M. McCloskey, and B. Green, "Naive Beliefs in 'Sophisticated' Subjects: Misconceptions about Trajectories of Objects," *Cognition* 9, 117—123 (1981).
- Students can often give correct answers to test questions without having accurate mental models.

C. R. Chamblis, "A Planetarium Oriented Sequence of Exercises," In *The Teaching of Astronomy*, J. M. Pasachoff and J. R. Percy (Eds.) (Cambridge University Press, Cambridge, U.K. 1990), pp. 387—388.
- The manipulation of celestial spheres and visits to interactive planetarium presentations improve student understanding of sky motion.

M. F. W. Dai, "Identification of Misconceptions about the Moon Held by Fifth and Sixth-Graders in Taiwan and an Application for Teaching." Dissertation Abstracts International (University Microfilms No. 9124300), (1991).
- Students think that moon phases are caused by Earth's shadow on Moon.

J. E. DeLaughter, S. Stein, C. A. Stein, and K. R. Bain, "Preconceptions abound among Students in an Introductory Earth Science Course," *AGU EOS Transactions*, 79(36), 429—432 (1998).
- Student-supplied-response exam shows students have difficulty describing a spherical Earth with a flat surface, that students have many contradictory conceptions, and that planetary motions are not understood. Data and test available on the Web at http://www.earth.nwu.edu/people/seth/Test (1998).

F. M. Goldberg and L. C. McDermott, "Student Difficulties in Understanding Image Formation by a Plane Mirror," *The Physics Teacher*, 24(8), 472—481 (1986) and F. A. Goldberg and L. C. McDermott, "An Investigation of Student Understanding of the Real Image Formed by a Converging Lens or Concave Mirror," *American Journal of Physics*, 55, 108—119 (1987).
- Exlploration of student difficulties with prediction strategies and semantics that physicists use when describing images, mirrors, and lenses.

D. Hestenes, M. Wells, and G. Swackhammer, "Force Concept Inventory," *The Physics Teacher*, 30(3) 141—158 (1992) and I. A. Halloun and D. Hestenes, "The Initial State of College Physics Students," *American Journal of Physics* 53, 1043—1056 (1985).
- Describes the rigorous development a nationally adopted conceptual test in physics education that is used to determine the effectiveness of various instructional techniques for students' conceptions of Newton's three laws of motion.

A. Lightman and P. Sadler, "Teacher Predictions Versus Actual Student Gains," *The Physics Teacher*, 31 (3), 162—167 (1993).
- Teachers predict much higher student achievement gains than students can actually achieve.

G. Mali and A. Howe, "A Development of Earth and Gravity Concepts among Nepali Children," *Science Education*, 63(5), 685—691 (1979).
- Students' conceptions of gravity are highly dependent on the specifics of their mental model of a spherical Earth.

National Research Council, National Science Education Standards (National Academy of Sciences Press, Washington, 1996).
- Describes a complete framework vision for a national content, instruction, assessment, program, and teacher professional development in science education. Available on the Web at http://www.nap.edu/books/readingroom/nses/.

J. Nussbaum, "Children's Conception of the Earth as a Cosmic Body: a Cross Age Study," *Science Education*, 63(1), 83—93 (1979).
- Students have a wide variety of mental models of Earth as a sphere.

J. Nussbaum and J. Novak, "An Assessment of Children's Concepts of the Earth Utilizing Structured Interviews," *Science Education*, 60(4), 685—691 (1976).
- Describes a highly effective and systematic interviewing strategy appropriate for investigating student conceptions.

R. J. Osborne and J. K. Gilbert, "A Method for Investigating Concept Understanding in Science," *European Journal of Science Education*, 2, 311—371 (1980).
- Describes a highly effective and systematic interviewing strategy appropriate for investigating student conceptions finding students have misconceptions about gravity.

W. C. Philips, "Earth Science Misconceptions," *The Science Teacher*, 58(2), 21—23 (1991).
- A long list of student conceptions in Earth and space science based on classroom experience of the authors.

G. Reed, "A Comparison of the Effectiveness of the Planetarium and the Classroom Chalkboard and Celestial Globe in the Teaching of Specific Astronomical Concepts," *School Science and Mathematics*, 72(5), 368—374 (1972).
- Students learn sky motions equally as well with a celestial sphere as in an interactive live planetarium teaching classroom.

E. R. Roettger, "Changing View of the Universe." Paper presented at American Association of Physics Teachers Meeting, New Orleans, LA. AAPT Announcer, 27(4), 4 (1998).
- Professional astronomers usually describe the universe as being very large and full of mostly empty space often starting with Earth and moving to larger entities. Alternatively, students usually don't have a holistic view of the Universe or are at ease to describe it.

M. M. Rollins, J. J. Dentton, and D. L. Janke, "Attainment of Selected Earth Science Concepts by Texas High School Seniors," *The Journal of Educational Research*, 77(2), 81—88 (1983) using methods described by D. A. Frayer, E. Schween-Ghatala, and H. J. Klausmeier, "Levels of Concept Mastery: Implications for Instruction," *Educational Technology*, 12(12), 23—29 (1972).
- Less than 80% of Texas high school students can accurately describe the cause of day and night.

P. Sadler, "The Initial Knowledge State of High School Astronomy Students," dissertation, Harvard School of Education Dissertation Abstracts International, 53(05), 1470A. (University Microfilms No. AAC-9228416)(1992).
- A comprehensive survey of high school student astronomy knowledge.

P. M. Sadler, "Psychometric Models of Student Conceptions in Science: Reconciling Qualitative Studies and Distractor-Driven Assessment Instruments." *Journal of Research in Science Teaching*, 35, 265—296 (1998) and P. Sadler, "Students' Astronomical Conceptions and How They Change," paper presented at American Association of Physics Teachers Meeting, Phoenix, AZ. AAPT Announcer, 26(4), 78 (1997).
- Students can move from one misconception to another misconception on their journey to scientifically accurate conceptions. Described using item response theory.

M. H. Schneps, *A Private Universe* (1987 videotape). Available from Pyramid Films and Video, 2801 Colorado Avenue, Santa Monica, CA 90404 (1987).
- Video of Harvard graduates and middle school students describing seasons and moon phases using faulty mental models.

P. S. Shaffer and L. C. McDermott, "Research as a Guide for Curriculum Development: An Example from Introductory Electricity," *American Journal of Physics*, 60(11), 994—1013 (1992).
- One of a series of seminal papers from physics education describing how educational research is used to create research-based curriculum materials.

C. Sneider and S. Pulos, "Children's Cosmographies: Understanding the Earth's Shape and Gravity," *Science Education*, 67(2), 205—221 (1983).
- Students think gravity is determined by proximity to Sun and rotation rate of planets.

K. Skam, "Determining misconceptions about astronomy," *Australian Science Teachers Journal*, 40(3), 63—67 (1994).
  • Students believe that Moon phases are caused by Earth's shadow.

T. F. Slater, "The Effectiveness of a Constructivist Epistemological Approach to the Astronomy Education of Elementary and Middle Level In-Service Teachers," Ph.D. dissertation, University of South Carolina (1993).
  • The same effective approaches used with children work well with in-service teachers to build confidence as well as knowledge. Extensive description of what constructivism looks like in astronomy.

T. F. Slater, J. R. Carpenter, and J. L. Safko, "A Constructivist Approach to Astronomy for Elementary School Teachers, " *Journal of Geoscience Education*, 44(5), p. 523 (1996).
  • Using hands-on activities founded in constructivism, elementary teachers significantly increased their astronomy knowledge and attitudes toward teaching it.

T. F. Slater, J. L. Safko, and J. R. Carpenter, "Long Term Sustainability of Teacher Attitudes in Astronomy." *Journal of Geoscience Education*. 47, 366—368 (1999).
  • Teachers who learned astronomy in a specially designed course maintained positive attitudes, values, and interests four years after the experience.

D. F. Treagust and C. L. Smith, "Secondary Students' Understanding of Gravity and the Motions of Planets," *School Science and Mathematics*, 89(5), 380—391 (1989).
  • Students think gravity is determined by proximity to Sun and rotation rate of planets.

S. Vosniadou, "Designing Curricula for Conceptual Restructuring: Lessons from the Study of Knowledge Acquisition in Astronomy." ERIC Document ED 404 098 (1992).
  • Students can easily accept that our Sun is hot but not that our Sun is a star because the "Sun as a star" concept is too far removed from direct experience. Similarly for day and night.

S. Vosniadou and W. F. Brewer, "The Concept of Earth's Shape: A Study of Conceptual Change in Childhood." ERIC Document ED 320 756 (1989).
  • Students' conceptions of a spherical Earth are highly distorted.

M. Zeilik, C. Schau, N. Mattern, S. Hall, K. W. Teague, and W. Bisard,
"Conceptual Astronomy: A Novel Model for Teaching Postsecondary
Science Courses," *American Journal of Physics*, 65(10), 987—996 (1997).

- Undergraduate courses using collaborative groups and concept
  maps erase many ethnic and gender differences that exist in
  pretest scores.

References

Adams, J. P., G. Brissenden, D. Duncan, and T. F. Slater (2001). "What Topics are Taught in ASTRO 101?" *The Physics Teacher*, **39**(1), pp. 8—11.

Adams, J. P. and T. F. Slater (1998). *Mysteries of the Sky*. Dubuque, IA: Kendall Hunt Publishing

Adams, J. P. and T. F. Slater (2000). "Stellar Bar Codes." *The Physics Teacher*, **38**(1), pp. 35—36.

Adams, J. P. and T. F. Slater (2002). "Observations of Student Behavior in Collaborative Learning Groups." *Astronomy Education Review (http://aer.noao.edu/)*, **1**(1).

Adams, Jeff and Tim Slater (2002). "Learning Through Sharing: Supplementing the Astronomy Lecture With Collaborative-Learning Group Activities." *Journal of College Science Teaching*, **31**(6), pp. 384—387.

Banks, J. A. (1999). *An Introduction to Multicultural Education* (2nd ed.). Boston: Allyn and Bacon.

Barab, S. A., K. E. Hay, M. Barnett, and T. Keating (2000). "Virtual Solar System Project: Building Understanding through Model Building." *Journal of Research in Science Teaching*, **37**(7), pp. 719—756.

Bianchini, J. A., D. J. Whitney, B. A. Hilton-Brown, and T. D. Breton (2002). "Toward Inclusive Science Education: University Scientists' Views of Students' Instructional Practices, and the Nature of Science." *Science Education*, **86**(1), pp. 42—78.

Bisard, W. and M. Zeilik (1998). "Restructuring a Class, Transforming the Professor: Conceptually Centered Astronomy with Actively Engaged Students." *Mercury*, **27**(4), pp. 16—19.

Bloom, B. S., Editor (1956). *Taxonomy of Educational Objectives: The Classification of Educational Goals: Handbook I, Cognitive Domain*. New York, Toronto: Longmans, Green.

Bonwell, C. and J. Eison (1991). *Active Learning: Creating Excitement in the Classroom (ASHE-ERIC Higher Education Report No. 1)*. Washington, DC: ASHE.

Brissenden, G., T. F. Slater, and R. Matheiu (2002). "The Role of Assessment in the Development of the College Introductory Astronomy Course: A 'How-to' Guide for Instructors." *Astronomy Education Review (http://aer.noao.edu/)*, **1**(1).

CAPER Team (2002). *Lecture-Tutorials for Introductory Astronomy*. Boston, MA: Prentice Hall.

Chickering, A. W. and Z. Gamson (1987). "Seven principles for good practice in undergraduate education." *AAHE Bulletin* **39**(7), pp. 3—7.

Comins, Neil (2001). *Heavenly Errors*. New York: Columbia University Press.

Deming G., and Beth Hufnagel (2001). "Who's Taking ASTRO 101?" *The Physics Teacher*, **39**(6), pp. 368—369.

Duncan, D. K. (1999). "What to Do in a Big Lecture Class Besides Lecture?" *Mercury*, **28**(1), pp. 14—17.

Francis, G., J. P. Adams, and E. J. Noonan (1998). "Do They Stay Fixed?" *The Physics Teacher*, **36**(8), pp. 488—490.

Hake, R. R. (1998). "Interactive-Engagement Versus Traditional Methods: A Six-Thousand-Student Survey of Mechanics Test Data for Introductory Physics Courses." *American Journal of Physics*, **66**(1), pp. 64—74.

Halloun, I. A. and D. Hestenes (1985). "The Initial State of College Physics Students." *American Journal of Physics*, **53**(11), pp. 1043—1055.

Hestenes, D., M. Wells, and G. Swackhammer (1992). "Force Concept Inventory." *The Physics Teacher*, **30**(3), pp. 141—158.

Hestenes, D., and M. Wells (1992). "A Mechanics Baseline Test." *The Physics Teacher*, **30**(3), pp. 159—166.

Hufnagel, B., T. F. Slater, G. Deming, J. P. Adams, R. Lindell-Adrian, C. Brick, and M. Zeilik (2000). "Pre-Course Results from the Astronomy Diagnostic Test." *Publications of the Astronomical Society of Australia*, **17**(2), pp. 152—155.

Lacey, T. and T. F. Slater (1999). "First Contact: Expectations of Beginning Astronomy Students." *Bulletin of the American Astronomical Society*, **31**(2). (Poster presented at the 194th meeting of the American Astronomical Society, Chicago, IL, 2 June 1999.)

Lyman, F. (1981). "The Responsive Classroom Discussion: The Inclusion of All Students." *Mainstreaming Digest*, University of Maryland, College Park, MD.

Mazur, E. (1997). *Peer Instruction: A User's Manual.* Upper Saddle River NJ: Prentice Hall.

Mestre, Jose (1991). "Learning and Instruction in Pre-College Physical Science." *Physics Today*, **44**(9), pp. 56—62.

Mestre, J. and J. Touger (1989). "Cognitive Research: What's In It for Physics Teachers?" *The Physics Teacher*, **27**(6), pp. 447—456.

Posner, G. J., K. A. Strike, P. W. Hewson, and W. A. Gertzog (1982). "Accommodation of a Scientific Conception: Toward a Theory of Conceptual Change." *Science Education*, **66**(2), pp. 211—227.

Redish, E. F., J. M. Saul, and R. N. Steinberg (1998). Student Expectations in Introductory Physics." *American Journal of Physics*, **66**(3), pp. 212—224.

Rosser, S. (1997). *Re-Engineering Female Friendly Science.* New York: Teachers College Press, Columbia University.

Rowe, Mary Budd (1974). "Wait-Time and Rewards as Instructional Variables, Their Influence on Language, Logic, and Fate Control: Part One—Wait-Time." *Journal of Research in Science Teaching*, **11**(2), pp. 81—94.

Seymour, E. and N. M. Hewitt (1997). *Talking About Leaving: Why Undergraduates Leave the Sciences*. New York: Westview Press.

Siebert, E. and W. McIntosh (2001). *College Pathways to the Science Education Standards*. Arlington, VA: NSTA Press.

Skala, C., T. F. Slater, and J. P. Adams (2000). "Qualitative Analysis of Collaborative Learning Groups in Large Enrollment Introductory Astronomy." *Publications of the Astronomical Society of Australia*, **17**(2), pp. 142—150.

Slater, T. F. and J. P. Adams (2002). "Mathematical Reasoning Over Arithmetic in Introductory Astronomy." *The Physics Teacher*, **40**(5), pp. 268—271.

Slavin, R. E. (1995). *Cooperative learning: Theory, Research, and Practice* (2nd ed.). Boston: Allyn & Bacon.

Tobias, S. (1994). *They're Not Dumb, They're Different: Stalking the Second Tier*. Tucson, AZ: Research Corporation.

Tobias, S and J. Raphael (1997). *The Hidden Curriculum—Faculty-Made Tests in Science, Part I: Lower-Division Courses*. New York: Perseus.

Tobias, S and J. Raphael (1997). *The hidden curriculum—faculty-made tests in science, Part 2: Upper-division courses*. New York: Perseus.

Tobin, K. and Capie, W. (1980). "The Effects of Teacher Wait Time and Questioning Quality on Middle School Science Achievement." *Journal of Research in Science Teaching*, I7, pp. 469—475.

Treisman, P. U. (1992). "Studying Students Studying Calculus: A Look at the Lives of Minority Mathematics Students in College." *The College Mathematics Journal*, **23**(5), p. 362.

Verner, C. and G. Dickinson (1967). "The Lecture: An Analysis and Review of Research." *Adult Education*, **17**, pp. 85—100.

Zeilik, M., C. Schau, N. Mattern, S. Hall, K. Teague, and W. Bisard (1997). "Conceptual Astronomy: A Novel Model for Teaching Postsecondary Science Courses." *American Journal of Physics*, **65**(10), pp. 987—996.